THE ILIAD
OF HOMER

To Don,

Happy Birthday,

[signature]

4/2/94.

THE ILIAD

OF HOMER

Retold by
**Barbara Leonie
Picard**

Illustrated by
**Joan
Kiddell-Monroe**

Oxford University Press

OXFORD NEW YORK TORONTO

Oxford University Press, Walton Street, Oxford OX2 6DP

Oxford New York Toronto
Delhi Bombay Calcutta Madras Karachi
Petaling Jaya Singapore Hong Kong Tokyo
Nairobi Dar es Salaam Cape Town
Melbourne Auckland

and associated companies in
Berlin Ibadan

Oxford is a trade mark of Oxford University Press

Copyright © Oxford University Press 1960
First published 1960
Reprinted 1960, 1961, 1966, 1975, 1980
First published in paperback 1991

A CIP catalogue record for this book is available
from the British Library

ISBN 0 19 274147 0

Printed and bound in Great Britain by
Biddles Ltd, Guildford and King's Lynn

PREFACE

HOMER'S two great epic poems, *The Iliad* and *The Odyssey*, both tell of happenings connected with the long war between the Greeks and the Trojans; but whereas *The Odyssey* is concerned with the adventures of one of the Greek heroes after the war was over, *The Iliad* relates incidents which took place in the ninth year of the siege of Troy: Achilles' quarrel with Agamemnon, his love for Patroclus, his friend, and how he avenged him.

Homer made his poems about three thousand years ago —at least four hundred years after Troy had fallen—and those for whom he made them would already have been familiar with the story and its characters. It was one of the great tales of their past, telling of the deeds of the heroes of the Greek people and of gods and goddesses they worshipped. Therefore, Homer wastes no time in giving the background to his story, but starts immediately with the quarrel between Achilles and Agamemnon, certain that his hearers will know both who they were and why they were besieging Troy.

It is different for us, three thousand years later, and so, in a prologue, I have set down the events which led up to the war of the Greeks with Troy; and, after Homer's tale of pride and love and vengeance is done, I have told briefly, in an epilogue, of the ending of the war and the fate of some of those who took part in it.

Barbara Leonie Picard

CONTENTS

PROLOGUE

How the War with Troy Began

IN a small kingdom in the north of Greece ruled Peleus, king of Phthia, who was much favoured by the gods. When he was no longer young, he fell in love with the sea-goddess Thetis, daughter of blue-haired and blue-bearded Nereus, who dwelt in the depths of the Aegean Sea, together with his consort, Doris, and their fifty daughters.

Lovely Thetis, though a lesser goddess, scorned the wooing of a mortal and fled from him, changing her shape in turn to that of fire, water, a serpent, and a fierce lioness. But Peleus was undaunted, and at last, with the help of Proteus, the Old Man of the Sea, he forced her to yield to his wooing, and she consented to be his bride: the only goddess ever to wed with a mortal man.

To the marriage feast of Peleus and Thetis, on Mount Pelion in Thessaly, came all the gods and goddesses, bringing gifts: two immortal horses, a spear, and golden armour. Eris, goddess of discord, alone was not bidden, for discord has no place amid the joys of a wedding.

But when the feasting was at its highest, Eris came among the feasters, and in her anger at the slight that had been put on her, she flung before them a golden apple inscribed with the words, '*For the fairest*'. Immediately, the goddesses began to vie, one with the other, for possession of the apple, and their strife continued in Olympus for a long time to come.

To Peleus and his immortal wife, a son was born, whom they named Achilles, and he early showed promise of courage and great beauty.

At a distance from Phthia, in the city of Opus, ruled Menoetius, who had a son named Patroclus, a little older than Achilles. While only a boy, by misfortune he killed another lad who quarrelled with him over a game of dice. On the advice of his father, Patroclus fled to Phthia, where King Peleus received him kindly and gave him to Achilles for a companion. From that moment a great love grew between them, and they were never willingly parted for even a single day. A few years later, when he was a youth, Patroclus, like all the young lords of Greece, became a suitor for the hand of Helen, daughter of Tyndareus, king of Sparta, who was famed for her beauty throughout all the land.

To the house of Tyndareus came, amongst many others, Menelaus, the brother of Agamemnon of Mycenae, the high king of Greece; Odysseus, king of Ithaca; Antilochus, the son of Nestor, king of Pylos; Diomedes of Tiryns and Sthenelus, his friend; Ajax, son of the king of Locris; Podaleirius and Machaon, the physicians; Idomeneus from Crete; Ajax and his half-brother Teucer, the sons of King Telamon of Salamis; and Patroclus.

King Tyndareus was greatly perturbed by the presence in his house of all these wooers, and he feared to choose a husband for his daughter from amongst them, lest he should anger all those others whom he rejected, bringing trouble on himself and on the man whom he had favoured; so he delayed his choice from day to day.

Odysseus, king of Ithaca, whose sharp eyes missed little, guessed his plight, and knowing that, as the king of only a small island, he himself stood slight chance of being chosen as Helen's husband, he spoke secretly with Tyndareus, bidding him let Helen make her own choice, after first asking from all the suitors an oath that they would abide by her choice and, furthermore, pledge themselves to give their help to her chosen husband, if ever at any time he should have need of help. To this plan Tyndareus thankfully agreed, and as a reward for his counsel, Odysseus asked for the hand of Penelope, the daughter of his brother, a maiden far less beautiful than her cousin Helen, but fair enough.

When all the suitors had sworn the oath, Helen made her choice; and she chose tawny-haired Menelaus, brother of great King Agamemnon. After the marriage feasting, the unsuccessful wooers returned to their own homes, Odysseus taking with him to Ithaca his bride Penelope. Soon after, Tyndareus died, and Menelaus succeeded him as king of Sparta.

Meanwhile, on Mount Olympus, where the gods dwelt, the quarrel still raged for possession of the golden apple, there being yet three rival goddesses who claimed it: Hera, wife of Zeus and queen of all the gods, Athene, goddess of wisdom and all crafts, and Aphrodite, goddess of love and beauty.

The gods could not decide between these three, so it was determined that a mortal man should be chosen to award the apple to that one whom he considered the most beautiful. Accordingly, the apple was entrusted to Paris, son of King Priam of Troy, and to him was left the judgement. He gave the apple to Aphrodite, who promised him, in return, that he should have for his wife the most beautiful woman in all the world.

The most beautiful woman in the world was Helen, and later, when Paris came to Sparta, she had been the wife of Menelaus for some three years. Menelaus received Paris kindly and made him welcome in his house, but while he was the guest of Menelaus, Paris won the love of Helen; and when he left Sparta to return once more to Troy, she went with him.

Menelaus appealed to his brother Agamemnon, and the high king at once sent word to Helen's former wooers, bidding them keep the oath that they had sworn. And they, with many other kings and lords of Greece, gathered together a great army and a fleet of ships, that they might sail to Troy and win back Helen for Menelaus and be revenged on Paris, who had so shamefully broken the sacred laws of hospitality by stealing the wife of his host.

Patroclus, as one of Helen's suitors, was amongst those called to go to war, and Peleus was willing that Achilles, though he was no more than a boy, should go with him and lead the Myrmidons, the men of Phthia, to battle, since Peleus himself was too old to fight.

But Thetis, as one of the immortals, knew that her son was doomed to the choice of a long and uneventful life, with no one to remember him when he was dead and gone, or a short life full of glory, and everlasting fame; and in an attempt to keep the choice from him, she clothed him like a girl and sent him secretly to King Lycomedes, on the island of Scyros, bidding the king hide him amongst his daughters. There, for a time, on Scyros, Achilles remained hidden, and there Deïdameia, one of the young daughters of Lycomedes, fell in love with him, and a son was born to them, who was later called Neoptolemus.

But the support of the brave Myrmidons, who were warriors of renown, was needed by the Greeks, and Odysseus went all about Greece seeking Achilles. Disguised as a merchant, he came to Scyros to the house of Lycomedes, suspecting Achilles to be there. He asked that he might show his wares to the king's daughters and their friends and handmaidens, and it was granted him. He spread out jewels and embroidered robes, girdles, and costly ornaments, whilst they crowded about him with cries of pleasure at the sight. He looked closely at them, yet he could not guess which amongst them might be a youth. But amid the jewels and trinkets, cunning Odysseus had laid a sword.

After a while, he made a sign to one of his followers, who went outside the king's house and there blew a war blast on a horn. With screams the maidens fled, or clasped their arms about each other, sobbing and trembling in their fear. But one tall maiden with golden hair, and beautiful above most, ran forward swiftly and snatched up the sword, and holding it and handling it as a sword should be held and handled, waited for the approach of danger. Odysseus smiled and called Achilles by his name, and Achilles did not deny the truth, for he wanted to go to Troy with Patroclus.

Achilles returned to Phthia, and his father was glad to see him once again, but Thetis grieved, for she saw that she could no longer keep his choice of doom from him. She told him what was ordained for him, and unhesitatingly he chose a short life and everlasting glory, and she warned him that he would never return from Troy. And so the Myrmidons, led by young Achilles and Patroclus, who left Phthia with Nestor of Pylos and Odysseus, set sail for Troy with the other kings and warriors of the Greeks, under the leadership of the high king, Agamemnon.

The great army of the Greeks encamped upon the shores of Troy, besieging King Priam's city. But if they were a mighty army, the Trojans, too, had many allies, and for a long time there was skirmishing, and fighting on the plain before the city; and sometimes the Greeks were successful, and sometimes the Trojans, but no decisive victory was won. And every now and then the Greeks would send off a certain number of their ships to sail along the coast from Troy, to sack one town or another in the neighbouring lands, both for cattle and plunder—for so great an army needed much food—and to discourage any further allies from sending help to the Trojans.

And so the war with Troy went on until it was in its ninth year.

I

The Quarrel

IN the ninth year of the war against Troy a great disaster came upon the Greeks.

A little time before, they had sacked a town, Lyrnessus, south-east along the coast from Troy. To young Achilles, son of King Peleus of Phthia, as the finest warrior amongst the Greeks, and to the Myrmidons, King Peleus' men, had fallen the greater part of the fighting, and they had taken much booty and many prisoners.

Amongst the captives was a young woman named Briseïs, and, when the fighting was done and the plunder gathered together, Achilles, as his share of the spoils, chose out Briseïs to be his slave.

Soon after this, Thebe, another town, close by Lyrnessus, was taken by the Greeks, and to the portion of Agamemnon, lord of Mycenae and high king of Greece, brother of

7

Menelaus for whose sake the war was being fought, and leader of the great army which had sailed to Troy, fell, as was customary, by far the largest portion of the spoils; and no one disputed his right. Amongst the captives whom he chose for himself was a fair maiden, Chryseïs, who was taken in his ship to the Greek camp on the Trojan coast.

Now, Chryseïs was the daughter of Chryses, high priest of the temple of Apollo, god of art and learning, and a man of riches and importance. Overcome with grief at the loss of his daughter, Chryses set sail from the ruins of Thebe with a great store of gold and silver, that he might ransom her from the Greeks.

On the coast of Troy, the Greeks had beached their ships in three lines along the low shore between two headlands; with the fifty ships of the Myrmidons farthest to the west, and those of tall Ajax, son of King Telamon, who had brought his father's men from the island of Salamis, at the extreme eastern point. Between the ships of Achilles and the ships of Ajax, the other ships were drawn up in three rows, for the stretch of level coastline between the high ground on either side was not long enough for a single line of ships.

The earliest to land had beached their ships in the first row, nearest to the fertile Trojan plain which stretched between the shore and the city of Troy, with the River Scamander winding through it.

Amongst the ships in this first row, and close to the middle of the line, were the forty ships which King Protesilaus had brought from Phylace. They were commanded by his brother, for Protesilaus was dead, in fulfilment of a prophecy which had declared that the first Greek to land on Trojan soil should be the first to die. Protesilaus, boldly leaping to shore while all the other Greeks held back, had been slain by Aeneas, a Trojan lord.

In the second row lay, amongst many others, the forty ships which had sailed from Locris under their leader Ajax, namesake and frequent battle-comrade of the son of Telamon, but, unlike him, short of stature, though valiant and hardy and skilled with a spear. His men, however, fought only with bows and slings, and carried neither spear nor shield.

In the third row, and nearest to the sea, lay the ships of the high king, Agamemnon, one hundred in all, by far the largest number to have sailed under any one leader; with, close beside them, the sixty ships which King Menelaus, his brother, Helen's husband, had brought from Sparta. In that row, also, were the ships of rich old Nestor, king of Pylos; and those eighty vessels with which brave young King Diomedes had sailed from his own land of Tiryns with his good friend Sthenelus. At the midmost point of this row were the twelve ships of Odysseus, king of the rocky little island of Ithaca to the west of Greece. Odysseus ruled over only a small kingdom, yet he was resourceful and skilled in using words to suit his ends, and he was much respected for his shrewd counsel.

In between the high hulls of the beached ships, raised up on stone supports to prevent their keels from rotting, ran narrow lanes, intersecting the two broad ways between the rows; and behind his own ships, each leader of the Greeks had built for himself and for his men, huts of wood and earth, well thatched with rushes cut from the banks of the Scamander, and shelters for their cattle and their slaves and stabling for their mules and horses, all with a palisade of stakes around.

In the very midst of the camp, before the ships of Odysseus, and stretching well into the second row of ships, lay the assembly place, where the kings and princes of the Greeks, together with their lords, would meet in council for judgement or discussion, and to offer sacrifice to the gods on the altar which stood at the centre of the open space. It was the right of any of the kings or princes to call an assembly to which all should come, and it was to this assembly place that Chryses was brought, to stand before the leaders of the Greeks, where they sat upon the benches, each in his wonted place, summoned by Agamemnon's heralds to receive the suppliant.

The old man stood before them, holding in his hands the sacred garlands from the temple and the sceptre of the god, and he pleaded with the Greeks that they would accept the ransom he had brought and give him back his daughter.

'Kings and lords of Greece,' he said, 'hear the entreaties of a father and restore my child to me, and so may divine Apollo look with favour on you, and all the other immortal gods who dwell on high Olympus, and grant that you may be victorious in your fight against the Trojans.'

All the kings and princes of the Greeks pitied him and cried out that it was fitting that a priest's words should be respected, that the ransom should be accepted and Chryseïs restored to him.

But Agamemnon, who had chosen Chryseïs for himself from amongst the other captives from the town of Thebe, said, 'The girl is mine, won by my spear, and it is my right to keep her if I wish. I say she shall not go. She shall stay here with me until Troy has fallen, and then she shall sail with me for Mycenae, and there in my palace, amongst my other slaves, she shall pass her days in spinning and weaving. Now go, old man, and do not come here again, lest another time you find the sacred garlands and the sceptre scant protection.'

The old priest was afraid, and he fled away along the shore while Agamemnon laughed; and because Agamemnon was the greatest king amongst them, and the leader of them all, no one disputed his decree.

But with tears old Chryses prayed to shining, immortal Apollo, 'O Bright One, I have served you well for many years. If ever I have pleased you, hear me now. Deliver my daughter from the hands of the Greeks, and avenge their slighting words.'

From far-off Mount Olympus, home of the immortal gods, Apollo heard him and, filled with divine wrath, he snatched up his quiver and his silver bow and came swiftly across sky and sea, and the sun was clouded by his mighty rage. He sat himself down at a distance from the camp of the Greeks and let fly amongst them his sharp arrows, winged with pestilence. At first he struck down mules and horses and hounds, and then he struck down men, until throughout the camp there was lamenting, and all along the shore the funeral pyres of the dead burnt ceaselessly.

For nine days did the plague rage, and on the tenth day, Achilles from Phthia, who was the youngest of all the kings and princes of the Greeks, called the other leaders to an assembly. When they were all gathered together, he stood up and spoke. 'My friends, without a doubt one of the gods is angry with us, else would this plague not have come upon us. Let us with no more delay ask a priest, or one who is skilled in such matters, to tell us which of the gods we have displeased by our neglect, that we may speedily offer sacrifices to appease his wrath. For otherwise, or so it seems to me, we shall never live to take Troy. We shall all of us die here, on the shore, and the theft of Helen will go unavenged.'

Achilles sat down, and immediately Calchas the prophet arose. 'Lord Achilles, son of King Peleus, I can tell you which of the gods is angered with us, and why. But first, swear to me that you will protect me, whatever I may say, for I fear that my words will displease one amongst us, who is the best of all the Greeks. Promise me your protection, and I will speak.'

'Have no fear, Calchas,' replied Achilles, 'but speak the truth to us, for while I live no man of all the Greeks shall harm you.' He paused a moment and glanced across to where Agamemnon sat, before adding, 'Not even great King Agamemnon himself, whom I have often heard declare himself to be the best of all the Greeks, shall do you ill, I promise it.'

'Then,' said Calchas, reassured, 'I will tell you why this plague has come upon us. It comes from immortal Apollo, and not because we have forgotten to sacrifice to him, but because King Agamemnon would not heed the pleas of old Chryses, who is Apollo's priest. Until the maiden Chryseïs is given back to her father, freely and unransomed, the god will not lift his wrath from us.'

Agamemnon frowned at his words, and, when Calchas had done, the great king sat silent, glowering, knowing that the eyes of everyone were upon him. Then he rose.

'Your words always make ill hearing, Calchas. Never have I heard you prophesy good fortune or the favour of the gods, but warnings and bad omens come always easily to your tongue. Now you choose to blame me for Apollo's wrath and this plague, and say that in my hands alone lies the remedy.' He paused, remembering how all the other kings and princes had been for accepting the ransom and restoring Chryseïs to her father. He flung out his arm in an angry gesture. 'Very well, the priest shall have his daughter back. I do not want to let her go, but I would rather lose her than be the death of all the Greeks. Odysseus shall make ready a ship and take the girl to Thebe.'

There was a murmur of approval from the assembled Greeks, but Agamemnon looked about him sullenly. He had been at fault and he had been set down before them all, and he was not minded to be the only one to suffer. He caught sight of Achilles, with his friend Patroclus sitting at his side. They were smiling at each other, now that it seemed likely that Apollo's wrath would soon be ended. Agamemnon thought how it was young Achilles who had called the assembly, and so caused his discomfiture. Suddenly he shouted out, 'I am the leader of you all, and the greatest king amongst you. It is not fitting that I alone should be without my chosen spoils. I shall send away Chryseïs because I must, but I will have instead a share of the booty of some other man—yours, perhaps, Achilles.'

Achilles sprang to his feet, his eyes blazing. 'You are our leader, King Agamemnon, but in this you presume too much. I have no quarrel with the Trojans. They have never stolen my father's cattle or his horses, nor have they harvested our fields of grain in Phthia. The mountains are

too high and the sea too wide that lie between Troy and
my father's land. No, I have led the Myrmidons to battle
against the Trojans for the sake of your brother's quarrel,
that he might win back his wife; though, since I was never
a suitor for Helen's hand—being too young by far—I was
bound by no oath to do so. Remember that, before you
talk shamelessly of sharing my booty.' He paused for a
moment, and then went on hotly, 'And what is my share
of the booty compared with all that you have taken from
the towns that we have sacked? Because I am accounted a
good warrior, and because my father's men are brave, to
me falls more than my fair share of the fighting, so long as
the battle lasts. But when the fighting is over, do I get a
like share of the spoils? No. After you, King Agamemnon,
have taken your pick, I get some small thing of little worth
from whatever is left over, and I have to be content with
that. What have I gained from this war with Troy? A
little gold and silver and a few slaves—not many, since,
unlike you, I am always ready to take a ransom for my
captives.' He paused again, and then said, 'By all the gods,
it would be better if I were to set sail with my father's men
and return to Phthia, rather than stay here, slighted, to win
more wealth for you to carry home to Mycenae.'

Agamemnon laughed. 'Go if you will,' he sneered. 'We
shall do well enough without you. You may be brave, and
a fine warrior, but of all the kings and princes of the
Greeks, you, with your arrogance and your hot temper
that you have not yet learnt to curb, you are the one whom
I could best do without. Go home. Take your ships and
your Myrmidons. Lord it over your own men, show your
temper to your friends and rail at them and not at me. Go

home; but part, at least, of your booty you shall leave
behind. When we sacked the city of Lyrnessus, you chose
as your share of the spoils the girl Briseïs, to be your slave.
I must give up Chryseïs to her father, since it is the will of
Apollo; but you shall give up Briseïs to me, since that is
my will.' He laughed again, shortly. 'And when she is
gone from your hut and dwells in mine, perhaps you will
have learnt at last that I am indeed the leader of all the
Greeks, and a greater king than your father Peleus.'

For a moment Achilles was too angry to speak or even
move, and then he laid his hand upon the silver hilt of his
sword, for in his rage he would have leapt upon Agamemnon and killed him there, in sight of all the other Greeks;
but before even Patroclus could restrain him, prudent
Athene, goddess of wisdom, who had been watching from
high Olympus, came swiftly down amongst them, unseen of any, and standing behind Achilles, she took hold of
his long yellow hair. He looked round and knew at once
that it was one of the immortal gods who held him back,
and he thrust his sword again into its scabbard, and immediately Athene left him.

Achilles, furious, but obedient to the divine injunction,
said, 'Your wits must be befuddled with wine, Agamemnon, that you speak so to one who is braver than yourself.
One day you will regret those words which have cost you
my help and the help of my father's men, for neither I nor
they, nor the friends who came with me to Troy, will fight
any longer against the Trojans for your sake.' He sat down,
flushed and still angry, beside Patroclus.

Agamemnon would have scoffed further at him, but old
Nestor rose. He was the king of Pylos, and had come with

ninety ships to fight against the Trojans with Antilochus his son, who had been one of Helen's suitors. He was the oldest of all the kings and princes of the Greeks, and his counsel was valued amongst them and his words were always heard with respect.

'Shame upon you both!' he exclaimed. 'How King Priam and his sons would rejoice to hear you now. I can chide you both, since I am older by far than either of you. In my time I have known greater warriors than any here today, and called them my comrades in battle. I fought beside Peirithoüs and great Theseus himself, against the Centaurs who dwelt in the mountain caves. No such grim fighting as that shall we see before the walls of Troy, yet we destroyed the Centaurs utterly. Even Theseus did not scorn my counsel, so why should you? Cease your quarrelling, my friends. Let Achilles keep the girl he won, King Agamemnon, she is his by right. And, as you have said yourself, he is a fine warrior, and the Trojans fear him. As for you, Achilles, you are yet very young; you should show more respect to a great king whose lands are far wider than any you will ever rule.'

'What you say is true, good Nestor,' said Agamemnon, 'but I am weary of Achilles' arrogance, his flouting of my authority, and his quick temper, which we see far too often.'

Before he could speak further, Achilles broke in, 'Your authority, King Agamemnon, do I no longer acknowledge. If you will, come and take the girl from me. I shall not prevent you. For I chose her by your favour: and what a man has given, that may he take away. Yet I warn you, lay but a finger on any of the goods I brought with me

from Phthia, and it will not be Trojan blood which will
be red upon my sword.' He rose abruptly, flung his cloak
about him, and gestured to his friend. 'Come, Patroclus.'
And without another glance at Agamemnon, he left the
assembly, and Patroclus went with him.

As Agamemnon had commanded, a ship was made
ready and, escorted by Odysseus, Chryseïs was sent to
Thebe, to her father; and the Greeks offered sacrifice to
Apollo, and he was appeased.

But Agamemnon called to him his heralds, Talthybius
and Eurybates, and said, 'Go to the hut of Achilles, son of
King Peleus, and take from him his captive, Briseïs. If he
resists you, come back and tell me, and I shall go myself to
fetch her and my warriors will go with me, and Achilles
shall soon regret his insolence and pride.'

Talthybius and Eurybates went unwillingly along the
shore towards the ships of the Myrmidons, for they had no
liking for the task they had been given. When they reached
the farthest westward point of the Greek camp, where
Achilles had his ships, they found Achilles sitting in the
porch before his hut, and at the sight of his angry frowns
they glanced at one another and stood silent, neither wish-
ing to be the first to tell him of Agamemnon's commands.

But Achilles looked at them and said, 'Greetings, heralds.
My anger is not against you, but against Agamemnon,
who has sent you. Yet, now that you have come, you can
be witnesses to my oath that never again will I fight in this
war against the Trojans for Agamemnon's sake.' He turned
away from the heralds. 'Fetch the girl, Patroclus, and give
her to them.'

Patroclus led out Briseïs from the hut and gave her to the

heralds, but she clung to him, weeping, for he had been kind to her on the day when she was taken captive, bidding her dry her tears and seeking to comfort her, saying, 'Achilles is young, and he has as yet no wife. You are nobly born, perhaps he may marry you and take you back to Phthia with him, where his father is a king, and you will have a fine marriage feast among the Myrmidons.' In tears again, now, she held on to his hand; yet there was no help for it, and at last she had to go with the heralds, walking with them along the shore towards Agamemnon's ships, looking back often through her tears.

But Achilles went alone down to the sea's edge, where the waves lapped upon the sandy shore and the rocks threw purple shadows on the water. And there, with tears of anger, he prayed to his mother Thetis that she might come to comfort him.

From where she sat, in the depths of the sea, in the halls of her father, blue-haired Nereus, Thetis, the immortal goddess, heard her son. She rose up from the water like a grey mist, and standing beside him, laid her hand upon his head.

'Mother,' he said, 'when I was called to Troy to lead my father's men you told me that the choice was ordained to me of a long and peaceful life, after which I should be forgotten of all men; or a short and glorious life, and after it, undying fame. As you know well, I chose a short life and a name that would live for ever in the hearts of brave men and in the songs of the minstrels. But of what avail was my choice, if in that short life I have to suffer the slights of such unworthy men as Agamemnon, son of Atreus? Oh, mother, you have told me that I have but a little while to live; soon, perhaps, may this brief life end before the walls of Troy. Where is the glory that was promised me?'

She stroked his hair and answered gently, 'My child, I would that you had never sailed to Troy. Yet here shall you win your undying fame and a place in brave men's hearts for ever. I will go myself to Father Zeus, and ask it of him.' And like a grey mist she faded from his sight, and he was alone again upon the shore.

Swiftly Thetis rose to the snow-crowned heights of Mount Olympus, to where Zeus, father of gods and men, sat upon his lofty throne, and there before him she fell down and clasped his knees in supplication.

'Father Zeus, there was once a time when all the gods rebelled against your rule, and I alone upheld you. If you are grateful for my loyalty, give me now the boon I ask of you for the sake of my son Achilles. So long as he holds himself from the war with Troy, by reason of his quarrel with Agamemnon, give the victory to the Trojans, that all the kings and princes of the Greeks may rue their leader's folly.'

For a time Zeus sat silent, deep in thought, seeing the things which were to come, and then he spoke gravely. 'I like it not, this thing that you have asked of me, for it will send many a good warrior, both Greek and Trojan, down to the land of my brother Hades, who rules the dead; and it will set all Olympus at strife, god warring with god and goddess against goddess, and all for the sake of your son. Yet since you have served me well, I will give you your boon, come of it what shall.' And he bent his head in token of assent.

II

The Combat

SOME twelve nights after, in fulfilment of his pledge,
Zeus sent a lying dream to Agamemnon, which
seemed to promise a great victory to the Greeks, and
soon. In the morning after the dream, Agamemnon told
the kings and princes of the Greeks how he believed that
victory was certain and the end of the long war in sight at
last, and he bade the whole army make ready for a mighty
assault on Troy.

Up from their camp came the army of the Greeks, men
and chariots—arms and armour flashing in the sunlight and
the horses neighing shrilly—like a great wave from the
shore, spilling on to the plain, leaving the camp empty
save for the slaves and the captives, and the few sick or
wounded; and—at the most western point—all those
who had come with Achilles from Phthia.

Each leader of the Greeks, in his light battle chariot, its strong frame made of wood and bronze with sides of wicker work and, beneath, a platform of plaited strips of hide to lessen the jolting, went at the head of his men, his charioteer standing at his side to guide the horses, and his chosen warriors running beside the chariot; and after him, in the first rank, came his lords and nobles in their chariots, with their own followers; and after them, the men who fought on foot. And so the mighty army of the Greeks moved forward over the plain, and only Achilles and Patroclus and the Myrmidons were not with them, for, since the quarrel, Achilles had not left his hut to fight, though such inaction irked him; and his men, too, at his command, remained about his ships, mending and sharpening their weapons and gear, dicing and playing games, practising their fighting skill or training their hounds, drinking, fetching fodder for their horses from the banks of the Scamander, and finding time go slowly for them.

The Trojans, wary and ever watchful, saw the Greeks approaching, and instantly made ready; and soon, out through the wide Scaean Gate, poured men and chariots to meet this new attack, far greater than any they had had to face before. Along the broad, paved way which led from the city, past the wild fig-tree that stood before the walls, past the burial mound of Ilus, one of the ancient kings of Troy, and on to the level plain, came the Trojans and their allies, led by Hector, the eldest son of King Priam and Hecuba his queen, and the greatest warrior in all Troy.

Close by the River Scamander, a good bowshot from each other, the two armies drew up in the midst of the plain, thick-starred with little red and yellow flowers; and there they waited, ready to attack, each leader turning to shout encouragement to his own men; while here and there a Greek or Trojan warrior would pick up one of the stones which lay upon the plain, and fling it, with a shouted taunt, in the direction of the enemy, a foretaste of what was yet to come.

But in that moment, whilst each army waited for the signal to attack, out from the Trojan ranks into the open space between the armies drove a single chariot, and from it there leapt down a young Trojan lord, handsome above most, with shining, richly ornamented armour and the spotted skin of a panther swinging from his shoulders: Paris, son of Priam and Hecuba, and the most hated of the Trojan princes, who had stolen Helen from her husband Menelaus.

Now, Paris, light-minded and careless, was not wont to court danger on the battlefield, and his favourite weapon was the bow, with which one might, if one were skilled enough, kill a man from a safe and sheltered place, well out of the reach of sword or spear. But on this day, being in a braggart mood, and not yet considering the consequences, he was rashly seeking that danger which he usually shunned. Brandishing two spears, he cried out a challenge to the bravest of the Greeks to meet with him in single combat.

Menelaus, seeing him, thought with satisfaction that at last he might be avenged on the man who had taken his wife from him, and instantly he leapt down from his chariot and came forward in acceptance of the challenge. Menelaus, king of Sparta, Agamemnon's brother, was by no means their greatest warrior, but he was brave and a good comrade in the fighting, ever ready to help others; and he was well liked amongst the Greeks.

But when Paris saw him there, sturdy and menacing, with his thick, reddish brows drawn together in a frown and his sandy beard glinting in the sun, he regretted his own rashness, and immediately stepped back into the Trojan ranks as though he would avoid a meeting with the husband of Queen Helen.

For all he was the son of their king, the Trojans murmured against him for faint-heartedness, and his brother Hector strode to him in anger. 'Most wretched Paris, would you shame us further? Have you not brought dishonour and grief enough on Troy already? Coward, and stealer of other men's wives, I wish that you had died before you went to Sparta. Would you have the Greeks make mock of us and say that we choose our champions for their good looks and not for courage? That I should have been cursed with such a brother! Of what use on a battlefield are your good looks, your winning ways, your curled hair, your skill at playing on the lyre? Fool that you are, you have made a challenge, and Menelaus has accepted it. Go out now, and meet the man whom you have wronged.'

Paris, white-faced, answered him, 'I will fight with Menelaus, if I must. But let it be for possession of Helen and the ending of the war. Let this meeting decide the outcome of the quarrel between our two peoples. Let him who wins take Helen, and let the loser renounce all claim to her.'

Hector, both surprised and pleased by his brother's answer, bade the Trojans hold from fighting whilst he spoke with the Greeks. He went alone between the armies and called to the Greeks for a hearing, and Agamemnon ordered them to be silent and listen to what Hector had to say.

Hector, speaking both to Greeks and Trojans, said, 'Hear the words of Lord Paris, son of King Priam, whose deeds brought this war upon our two peoples. He will fight alone with King Menelaus, son of Atreus, for possession of Queen Helen, if every man here will swear to respect the outcome of the battle, and let the winning of Helen end the war.'

When he had finished, no one spoke for a time, then Menelaus glanced at Agamemnon, who gave his brother a nod of agreement, and Menelaus stepped forward and said, 'I agree to the conditions, but let King Priam himself come from the city and swear to them in the name of all the Trojans and their allies.'

Both Greeks and Trojans were glad when they saw the ending of the long war so close. They laid aside their weapons and took off helmets and breastplates and sat on the ground to wait, while Hector sent two heralds running to the city to summon Priam.

In Troy, Helen, in the house of Paris, sat embroidering on linen the tale of the enmity between the Trojans and the Greeks. When she heard of the challenge of Paris and how Menelaus had accepted it, she threw a veil over her lovely hair, and calling two of her handmaidens to go with her, she went out from the house, which stood close by the palace of King Priam and the house of Hector, on the high citadel called the Pergamus, in the very midst of the city. Down the wide street she went, which led to the Scaean Gate, beside which rose a tower from where one might watch what went on upon the plain.

She found King Priam and his counsellors sitting in the watch-tower, old men all of them, too old to fight, but respected for their wisdom and good advice. They looked up and saw her coming. 'It is Helen,' they said, and in low voices they spoke amongst themselves. 'It is small wonder that the Trojans and the Greeks fight over her, for she is as fair to look upon as one of the immortal goddesses. Yet for all that, I wish she were gone from Troy, for she has brought much grief to us, young and old alike.'

But when Priam saw her, he smiled and made room for her close by him on the stone bench where he sat, for he had always shown her courtesy since the day when Paris had first brought her home to Troy. 'Come, my dear child,' he said, 'and sit with us and point out to us the kings and princes of the Greeks, for they will all be known to you by sight.'

So Helen pointed them out: Agamemnon, his breast-plate ornamented with blue enamelling, Odysseus with the dark-red hair, tall Ajax, son of Telamon, carrying the huge shield, young Diomedes with Sthenelus, grey-bearded old Nestor, whose shield was solid gold, little, quick Ajax of Locris, and handsome Idomeneus, the king of Crete; and King Priam and his counsellors praised them for goodly seeming men. Yet the youngest and most beautiful of all the kings and princes of the Greeks was not amongst them. Achilles, on the shore, had climbed upon a ship, and standing at the high stern, narrowing his eyes against the morning sun, while the wind blew through his golden hair, he was looking out across the plain, to see what he could of all that was taking place.

While Priam and Helen and the old men sat talking in the watch-tower, old Idaeus, Priam's herald, came to tell the king of how two messengers were come from the battlefield with word of the single combat between Paris and Menelaus which was to decide the outcome of the war, and of how Priam himself was called upon to take an oath in the name of all his people.

Bidding wise old Lord Antenor, his chief counsellor, to go with him, Priam went down from the watch-tower to his chariot, awaiting him before the Scaean Gate. Followed by the heralds bearing wine and offerings for a sacrifice, he drove swiftly to where the two armies waited, and there, in the open space between them, he came down from the chariot and walked towards the Greeks, with Antenor at his side.

Immediately, Agamemnon rose, beckoning to Odysseus, and they went forward to meet the Trojan king, while the heralds brought the offerings. Then, when the two kings, Priam and Agamemnon, had cleansed their hands in water, they poured out the libation and sacrificed three sheep, and took an oath to end the war and abide by the outcome of the single combat.

Then Greeks and Trojans alike made ready eagerly to watch the combat, but old Priam said, 'I will go back to the city, for I have no heart to stay and watch my own son fight. Immortal Zeus will give the victory as he sees fittest.' And, with Antenor, he mounted into the chariot and returned to Troy, sad at heart for his son, guilty and worthless though he might be.

Hector and Odysseus measured out a space between the armies and cast lots from a helmet to see which of the two, Paris or Menelaus, should be the first to hurl his spear at the other; and the lot fell to Paris to throw first.

Menelaus and Paris armed themselves, Paris putting on the breastplate of his half-brother Lycaon, for it was stronger than his own, though far less richly ornamented. He had taken it for himself at the time when Lycaon had fallen captive to Achilles, who had sold him into slavery across the sea.

Then the two of them stepped forward and stood opposite each other in the space which had been measured out, and, since the lot had fallen to him, Paris was the first to cast his spear, and it struck the shield of Menelaus. The shield was made of layers of stout hide, with an outer layer of bronze, and the bronze held, and turned the point of the spear.

With a prayer to Zeus, 'Immortal Zeus, father of gods and men, protector of guest and host, grant I may now be avenged on Paris, who did me such great wrong; so that, in years to come, my vengeance may be remembered and men shrink from doing evil deeds in the house of a kindly host,' Menelaus hurled his spear, and with such a mighty effort, that it passed right through the shield of Paris and on through the edge of Lycaon's breastplate, even tearing the linen tunic which Paris wore beneath, but leaving him unharmed: yet by no more than a hair's-breadth had Paris escaped from death.

Menelaus drew his sword, and raising it, ran forward and struck him a mighty blow upon the helmet; but the

hard bronze ridge of the helmet, to which the crest was
fixed, broke the sword into four pieces, which fell upon
the ground. 'Zeus, Zeus,' cried Menelaus, 'do you deny me
my revenge?' But he was not daunted, and, weaponless,
he flung himself on Paris and, seizing him by the crest of
horsehair on his helmet, threw him down and would have
dragged him towards the Greek lines. Paris, almost choked
by the leather strap which fastened the helmet beneath his
chin, could do nothing to help himself; but the strap broke
and the helmet came away in Menelaus' hand. He flung
it from him and leapt again upon Paris, meaning to catch
him by the hair and hold him fast until he could take up
his spear from the ground.

But golden, laughing Aphrodite, goddess of love and
beauty, whom Paris had adjudged the loveliest of all the
immortal goddesses, would not permit him to be slain,
and in a cloud she gathered him up, carrying him in an
instant to his own house. There she set him down, astonished, but thankful indeed that he had come alive from his
battle with Menelaus.

Menelaus, baffled, saw Paris vanish from before his very
eyes, and catching up his spear, he sought him furiously
up and down amongst the Trojan lines, thinking that perchance some comrade was protecting him. But indeed, few
Trojans would have hidden him from Menelaus, for he
was too well hated for all the grief that he had brought to
Troy.

When Helen, who had watched the battle from the
watch-tower beside the Scaean Gate, learnt from Aphrodite
that Paris awaited her in his house, she could not at first

believe it, for she had seen him, only a little while before, being worsted in his fight with Menelaus.

But she knew that if Paris were indeed safe in his own house, it could only be through the protection of the immortal gods, and she made haste to go to him up the street that led to the high Pergamus. Yet she went silently and thoughtfully, for she had that day looked once again on Menelaus, her generous, trusting husband, whom she had loved and chosen from amongst all her many wooers; and seeing him once again, bold and sturdy, a man worthy of regard, she had bitterly regretted ever leaving him.

She found Paris in his own chamber, smiling and well pleased, looking more like one who had just come from a dancing ground than from a battlefield.

She sat down, but would not look at him. 'So,' she said, 'today you have fled from a better man than yourself. How often you have boasted to me in the past of all that you would do if you met with Menelaus in the fighting! Having slipped safely through his hands today, will you dare challenge him again? If you do, you will be a fool, for he is by far the better man.'

He went to her, gaily, tossing back his dark curls. 'Come, Helen, do not reproach me for it, that the gods were with Menelaus today. Another time, who knows, they may give me the victory.' He took her hand, coaxingly, with fond looks at her lovely, frowning face. 'Be kind, Helen, and smile at me again, for I still love you. Indeed, I think that I now love you even more than on that first day when you came so willingly with me from the house of Menelaus.'

III

The Broken Truce

So Paris spoke with Helen, and smiled; but Menelaus, having searched the plain and found no trace of him, returned to his brother Agamemnon, who said, 'Whether false Paris has hidden himself or has been hidden by one of the immortal gods—or, indeed, whether he is alive or dead—is of no consequence. What matters is that, before the eyes of all, you worsted him. Therefore let the Trojans keep the oath, sworn by King Priam in their name, and return Helen to you, my brother, so that there may be an ending of this war.'

And so indeed might they have done—for Trojans and Greeks alike were wearied of the years'-long fighting—but that Zeus, watching from his everlasting halls on Mount Olympus, remembered his promise to Thetis, and calling to him his daughter Athene, who gave to men all skill and craft, whether of hands or mind, he bade her go down to earth and put it into the mind of one of the Trojans to violate the truce.

Athene, flashing-eyed and sternly beautiful, answered him, 'It shall be as you say, Father Zeus.' She took up her tasselled shield, terrible to look upon, and wearing her helmet of imperishable gold, immediately she departed from Olympus, gleaming like a star. Well pleased was she to obey the command of Zeus, for she had been ill disposed towards Troy since Paris had awarded the golden apple of Eris to her rival, Aphrodite, and she had no wish to see the war between the Greeks and the Trojans ended until all Troy had paid for Paris's offence, and Priam's city lay in ruins.

On the plain before the city, in the likeness of a young Trojan nobleman, she sought out Pandarus, one of the allies of the Trojans, who had come to the aid of King Priam from Zeleia, at the very foot of the rugged, pine-clad slopes of Mount Ida, south-east of Troy, and she spoke to him with enticing words. 'Brave Pandarus,' she said, 'does it not seem to you, high hearted as you are, a shameful thing that the champion of the Trojans should have been overcome by Menelaus? Should any Greek live to boast of a victory over one of great King Priam's sons? Think you not it would be a good thing, and one which would win you much honour amongst the Trojans, if Menelaus were to fall to an arrow from your bow? Like Prince Paris, you are skilled with the bow, rarely do you miss your mark, as we of Troy well know. Come, my friend, there stands Menelaus, well within reach of an arrow; do not hesitate. Kill Menelaus, and you will win, not only much honour from the Trojans, but, as well, rich gifts and much wealth from Prince Paris. For he will be grateful to the man who is bold and skilled enough to rid him of Queen Helen's lord.'

Flattered by her praise and seduced by her subtle words, in his folly, Pandarus took up his bow, fashioned of the horns of a wild goat which he had slain on Ida, and fitted to it an iron-tipped arrow. Before any standing near him could have prevented him, he had taken aim at Menelaus, with a prayer to Apollo the Archer on his lips.

Swiftly the arrow flew, but Athene was swifter, standing before Menelaus, unseen of any, to turn the arrow from its mark. To his belt of leather set with gold she guided it, and it passed through the belt and pierced the flesh beneath. But the golden clasps of the belt stayed its course, so that no more than the tip of the arrow entered his side, though the blood flowed down, reddening his tunic.

Agamemnon, standing beside him, exclaimed in horror, and put his arms about his brother. 'Menelaus, best of brothers, I have brought you to your death by rashly trusting in the good faith of our enemies,' he cried. 'Ever after will it be said of me that through my folly my brother died. When the war is over and the Greeks sail home again, you will be left alone in alien ground for the Trojans to dishonour your bones, leaping vauntingly upon your burial mound to boast, "Agamemnon is gone home, yet he came to Troy in vain, for he has left Menelaus behind him."'

But Menelaus made light of his wound, saying cheerfully, 'It is no more than a scratch. See, how the barbs of the arrow have caught on the clasps of my belt. There is no need yet to raise the dirge for me. And calm yourself, I beg of you, good brother, or we shall have my men thinking that I am half-way to Hades' land.'

Yet Agamemnon knew too well his brother's courage and his usual bluff manner of speech, to be entirely reassured. 'May the immortal gods grant that you speak truly,' he said. 'But I shall not be satisfied until Machaon, our good physician, has seen the wound and told me so himself.' And he called out to Talthybius the herald to fetch Machaon to them, and, smiling, Menelaus shrugged his shoulders and let him have his way.

When Machaon was come, he cleaned the blood from the wound and laid healing herbs on it, binding it with strips of linen; while Menelaus jested with him.

Once Agamemnon was reassured that his brother was indeed not badly hurt, he turned his thoughts to the Trojans. 'Now that they have broken their oath,' he said, 'surely they will be justly hated of the gods, and victory shall be ours. We cannot fail to destroy them now.' And forthwith he commanded the Greeks to make ready for instant battle, going himself to each leader, urging him to fight his utmost to avenge the broken truce.

First he went to the Cretans, where they were arming themselves eagerly, impatient to take vengeance on the Trojans for their treacherous oath-breaking. Idomeneus, their king, stood before them, calling out to encourage them; while young Meriones, his friend, exhorted the stragglers at the rear. Idomeneus was one of the older leaders of the Greeks, his hair was already a little grizzled, but he was still as handsome as ever he had been, as upright and as strong; and he was, moreover, a rich and important ruler.

Agamemnon, gladdened by the eagerness of the Cretans, praised King Idomeneus for his readiness and thanked him for his support, before passing on to where Ajax of Locris and Ajax, son of Telamon, prepared for battle together, with the Locrians and the men of Salamis, well armed and ready, standing near their leaders.

'You are valiant men, both of you,' said Agamemnon, 'and you need no encouragement from me. If all the Greeks had your daring spirit we should take Troy this very day.'

He found old King Nestor ready armed, his gold shield flashing in the sunlight, as he drew up the men of Pylos in their ranks, ordering the charioteers to keep close together on the field, in one unbroken line, each chariot lending support to the others, and not to fight, each man, selfishly, for his own glory, pressing forward alone against the enemy.

'Good Nestor,' said Agamemnon, 'I wish that your strength were as ageless as your wisdom, or that your old age might leave you and fall instead upon some younger man, giving you his youth, for there is no one of all the kings of Greece like you.'

'Old I may be, great King Agamemnon,' replied Nestor, 'but I can still strike a blow against the Trojans, who have shamefully broken their oath, as they shall learn today. For the rest, I can urge younger men to fight bravely, and set them an example.'

And so Agamemnon went from one leader to another, praising and encouraging, well pleased by the zeal he found. But when he came to those who were farthest from his own men of Mycenae—King Odysseus with his followers from Ithaca, few but brave; and the men of high-walled Tiryns, under young King Diomedes—he found them unarmed and idle, for they had not heard his call to battle, being too far away.

Frowning, Agamemnon spoke to Odysseus. 'Why do you stand here avoiding battle and waiting for other men to do your fighting for you? Full of craft and guile as you are, Odysseus, doubtless you will say that you did not hear me call all men to battle. Yet when I summon the Greeks to a feasting, you never fail to hear me, however far off you may be.'

Odysseus was angry, as well he might be, at being so unjustly taken to task, and he did not hesitate to retort to Agamemnon's words. 'Are you calling me a coward, King Agamemnon, because I did not hear your command? Have you not often seen me in the front of battle? Yet perhaps you were not there yourself to see me.' He made a scornful gesture. 'One should always think well before one speaks, lest one speak foolishly.'

Seeing that he had angered Odysseus by his ill-considered words, and being unwilling to lose the support of another leader of the Greeks, Agamemnon said hastily, 'I do not doubt your courage, Odysseus, nor the courage of your good men. And I am certain that I do not need to urge you to battle.' He smiled ingratiatingly. 'If I have spoken mistakenly, you must forgive me, for I meant no slight.'

Hurriedly he passed on to the men from Tiryns, and, irritated that he had been put out of countenance in his difference with Odysseus, he began immediately to upbraid Diomedes, calling out loudly to him as he approached, 'I have heard, Diomedes, that your father, Tydeus, was a brave man, and one who was ever found in the front of battle. It seems to me that Tydeus' son lacks his father's courage.'

Diomedes, mastering his anger, bit his lip and did not answer, for he was a younger man than Agamemnon and a far lesser king, and would have shown him respect; but Sthenelus, who was at his side, would not stand by and hear his friend insulted, even by the high king of Greece. Flushed and indignant, he stepped forward. 'That is a lie, King Agamemnon, and well you know it. Diomedes is no coward.'

But before he could say further, Diomedes turned on him, speaking, in his anger with Agamemnon, far more harshly than he meant. 'Hold your peace, Sthenelus. It is King Agamemnon's right, as the leader of us all, to rebuke me if he sees fit. For he commands us, and if we take Troy the glory will be his; but if we are defeated, then the shame, also, will be his. So he is both wise and in his rights to exhort and chide us. Come, let us heed his reproach, Sthenelus, and go forth against the Trojans.' And without a word to Agamemnon, he turned away and made ready to go into battle, that he might wash out, in Trojan blood, the sting of Agamemnon's injustice.

When battle was joined, there were many Greeks fought well that day, yet none so fiercely as Diomedes; and few Trojans who came within reach of his spear lived to tell of it.

But while he raged across the plain, and the Trojans fled

from him in terror, Pandarus from Zeleia, who had broken the truce by wounding Menelaus, seeing him come within range of his bow, took aim at him. Straight sped the arrow, and its barbed iron head passed clean through Diomedes' brazen breastplate and on into his shoulder, so that Pandarus, seeing it, cried out in triumph to his comrades, 'Take courage, men of Troy, for I have slain Diomedes, who is the most fearful of those who stand against us today.'

Diomedes withdrew from the fighting to where Sthenelus waited with his chariot and horses, for Diomedes had left him, very early in the battle, to fight on foot. He called, 'Sthenelus, I am wounded. Come to me.' And Sthenelus, hastily tying the reins to the front rail of the chariot, leapt down and ran to him, much distressed.

'Pull out the arrow,' said Diomedes.

Sthenelus set his teeth and drew the arrow out, and the blood streamed down Diomedes' breastplate. Sthenelus staunched the wound as well as he could.

'Can you hear that braggart boasting how he has slain me?' asked Diomedes. 'Not long shall he live to vaunt, if the gods are kind to me.' He prayed aloud to Athene, 'Immortal daughter of Zeus, if ever you stood by my father, Tydeus, in battle, stand by me today. Let Pandarus come within reach of my spear, and grant that I slay him.' And with that, though Sthenelus would have prevented him, he returned into the thickest of the fighting, to seek out Pandarus; while Sthenelus drove the chariot and horses after him, keeping as close as he could, unwilling to lose sight of Diomedes for an instant, lest he neeeded help.

Aeneas, son of a great Trojan lord, and the bravest

warrior of the Trojans after Hector, saw him from his chariot and called out to Pandarus, who was near, 'Here comes Diomedes, son of Tydeus, bent on further slaughter. Loose an arrow at him, good Pandarus, and save many Trojan lives.'

Pandarus, astonished and disconcerted, exclaimed, 'Does he still live? My arrow struck him through the breastplate but a short while ago. Truly, the gods have granted me ill fortune today. Two kings of the Greeks have my arrows reached, yet neither have I sent to his death. There is no luck on my bow. I wish I had not brought it with me. A chariot and horses would have served me better. In my father's stables stand eleven new-made chariots, and a pair of horses to each. He bade me choose from them before I came to Troy. Yet I thought that, with Troy besieged, it might not be easy to find fodder for the horses, so I trusted to my bow instead and came as my men, on foot. If I live to see my home and my wife again, may some man strike the head from my shoulders if I do not break this wretched bow of mine with my own hands and cast it into the fire.'

'Come, do not be downhearted, my friend,' said Aeneas. 'Mount beside me in my chariot and together we will go against Diomedes. And if he should prove to be too mighty for the two of us—which I doubt he could be, unless a god stands by him—why then, my horses are swift and will bear us safely from him, for they are of the same strain as King Priam's horses, which are the best in Troy. Come now, take the reins and whip and I will fight; or, if it is more to your mind, let me drive whilst you attack him.'

Eagerly Pandarus leapt into the chariot. 'The horses will answer better to your guidance, noble Aeneas, for they know your hand and voice. Let you drive, and I will strike at Diomedes with this good sharp spear of mine. It were best we did not risk the horses' running wild at a moment of danger, because a strange hand guided them.'

Aeneas urged the horses on, and together they made towards Diomedes. Sthenelus, close behind his friend, saw them coming and warned Diomedes. 'Here comes Pandarus against you once more. But this time he is with Aeneas. You are tired and you are wounded, and you cannot face the two of them. Come into the chariot and I will take you out of their reach.'

Diomedes' shoulder pained him, and Agamemnon's words still rankled, and he answered sharply, 'I am no coward to fly from battle even if it is you who bid me. I shall meet those two as I am, on foot.' He added, after a moment, in better humour, and with an eagerness hardly hidden by his too casual manner, 'If I should chance to kill them both, Sthenelus, be ready to take the horses of Aeneas, for they are among the best in Troy, it is said.'

As they neared each other, Pandarus leant from the chariot crying out, 'Son of Tydeus, my arrow did not kill you, may my spear have better fortune,' and hurled his spear at Diomedes. The spear passed through Diomedes' shield, and touched his breastplate; but its force was by then spent, and it did not pierce the armour.

Pandarus shouted in triumph, for he saw the point of the spear pass through the shield and he thought that it had passed on into Diomedes' body; but, even as he shouted, Diomedes' spear struck his head and he fell from the chariot. Aeneas, dropping the reins, leapt down after him, and stood with spear and shield to defend his body, lest

Diomedes should seek to strip the armour from it. But Diomedes remained where he was, and taking up a huge stone which lay on the ground at his feet, he flung it at Aeneas with such force that Aeneas fell to the ground, and everything went dark before his eyes. Diomedes ran forward, sword ready, to take his life, but golden Aphrodite, unseen, spread a fold of her bright robe over him, and he vanished from Diomedes' sight.

And so, through the intervention of immortal Aphrodite, Diomedes failed to kill Aeneas; but Sthenelus, leaping swiftly from the chariot, left his own horses to one of Diomedes' followers, and took the horses of Aeneas as spoils for his friend. And very glad were Diomedes and he to have won so fine a prize.

Across all the plain of Troy the battle raged, with Athene lending strength and courage to the Greeks, and Aphrodite ever ready to protect the Trojans. And Ares himself, dread god of war, came down from Olympus to the battlefield. To him it mattered not who lost or won, so long as men fought and died; but on this day he fought beside the Trojans, for he was the lover of Aphrodite, and she wished the Trojans well.

When Ares was come against them, the Greeks gave ground and retreated, though slowly, and fighting all the way, towards the ships; until at last, in wrath, Athene drove Ares back to Olympus and, for a time, the gods ceased warring amongst mortal men.

With Ares gone from the field, the Greeks took heart and rallied, and the Trojans fared ill at their hands. Ajax, son of Telamon, fought mightily, sparing none, as did

Antilochus, Nestor's son; while Nestor himself forbade his men to cease fighting long enough to strip the slain. 'When all our enemies lie dead,' he said, 'there will then be a time for the taking of spoils.'

Yet, even in the midst of slaughter, there were moments of pity and forbearance. As Adrastus, a rich young Trojan, was fleeing from Menelaus, the wheels of his chariot became entangled in the branches of a tamarisk bush, the chariot pole snapped and Adrastus was thrown to the ground.

Menelaus leapt from his own chariot and came up to him, spear in hand, and Adrastus clasped his knees in supplication, begging Menelaus to take him alive. 'My father has much treasure laid by in his house, good son of Atreus. He will pay you a great price for my life,' he pleaded.

Menelaus hesitated, then, pitying him, turned to call his followers to bind the youth and take him to the ships; but Agamemnon, seeing his brother about to spare a suppliant, came running to him. 'Why so weak and soft-hearted, Menelaus? Have the Trojans done you so much kindness that you should take thought for them? Let us slay them all. Let us not leave one of them alive in Troy.'

Though he was his own brother, Agamemnon was also high king of Greece and leader of all the Greeks at Troy, so Menelaus obeyed him. He unclasped Adrastus' hands from about his knees and stepped aside; yet he did no more than that, and it was Agamemnon who thrust his spear into the young man's body.

More happy was the meeting between Diomedes and

Glaucus the Lycian, who had come to Troy to fight for King Priam with Sarpedon, his friend and fellow-king. The Lycians were a folk who lived far away from Troy, in the south, and were strangely clad, in ungirdled tunics. Yet they were the most respected of all the Trojan allies, and their two brave kings were much honoured.

When all who could were keeping as far from Diomedes as they might, so much were all the Trojans dreading him that day, Glaucus came forth alone and challenged him.

'Who are you, brave stranger, who come so willingly to die at my hands?' asked Diomedes.

'Diomedes, son of Tydeus, why do you ask my name? What matters who I am? Even as leaves upon the trees of the forest are the generations of men. The leaves fall in the autumn and the wind blows them away; and in the spring the trees put forth new leaves. Even so perishes one generation of men, and is forgotten; but another comes to take its place. Yet, since you ask, my father was Hippolochus, and his father was Bellerophon, and I am Glaucus, who rules in Lycia with King Sarpedon.'

Immediately Diomedes thrust the point of his spear into the ground, and leaving it standing so fixed, went towards Glaucus with a smile, saying, 'Then should we be friends, for Oeneus, my father's father, once entertained Bellerophon in his house, and they exchanged gifts. My father Tydeus I do not remember, for he died when I was very young, but Oeneus has often told me of Bellerophon and shown me the golden cup he gave him. So we, too, should be friends, even as they were; and if ever, in happier days, you chance to come to Tiryns, you shall be my guest, and I yours, if ever I am in Lycia. Let us exchange gifts, as our grandsires did, and keep from each other in the fighting. For surely there are Trojans and their allies enough for me to kill, and Greeks enough for you to slay, without our taking each other's lives.'

So they clasped hands in friendship and pledged faith with each other, and exchanged their armour as a gift. And though the armour of Glaucus was all of gold, while Diomedes' armour was only bronze, and a poor exchange, yet were they both well satisfied, and gladdened by their encounter.

IV

The Unaccepted Offering

WHILE Hector fought, striving with all his might to resist the Greek advance, Helenus, who was a younger son of Priam and Hecuba—no great warrior, yet a young man of good sense and considered purpose—came to him and said, 'Things go ill with us, brother. Should we not make offerings to the immortal gods, and pray for help? Above all, should we not make an offering to divine Athene, that she may turn from the Greeks and give us her aid? Whilst we others remain here and hold back the enemy, let you, Hector, return to the city and ask our mother to call all the women of Troy to the temple of Athene, and there make her an offering. Then perchance she may spare us from Diomedes' attack, for he seems as mighty today, and as much to be feared, as ever Achilles was, in the days before he held from fighting.'

Hector, more used to fighting and to leading men than to devising ways and means, was always impressed by his young brother's quick understanding, and ready to be guided by him in such matters. 'That is well counselled, Helenus,' he said. 'I shall do as you advise.' He leapt down from his chariot and went amongst the Trojans, encouraging them and bidding them fight bravely while he was gone; then, slinging his shield behind his back, he ran swiftly towards the city.

As he neared the great oak-tree, sacred to Zeus, which stood before the Scaean Gate, there came hurrying to meet him along the broad, paved way, a number of women who had been waiting beneath the gate: the wives and mothers of some of those who had gone out to fight that morning, when more of their enemies than ever before, at any one time, had been sighted coming up from the shore against the city. Anxiously the women asked him news of their menfolk: how this one did, whether that one were unhurt, and if he had seen some other.

'Pray to the gods for them, and for all in Troy,' he said, and would not stop to answer further, but ran on through the gateway.

He climbed the wide street to the Pergamus, where stood the palace of Priam, the houses of his sons, and the temples of the gods, and there, at the doorway of the palace, his mother, Queen Hecuba, met him. She took his hands in hers and asked him apprehensively, 'Why have you come from the battle, my son, and alone? Is it to ask help of the immortal gods?' Hector nodded, but before he could speak and answer her, she said, 'First let me fetch you a cup of wine, that, when you have poured a libation to the gods, you may drink and be refreshed.'

But he shook his head and laid a hand upon her arm when she would have gone from him. 'Fetch me no wine, dear mother, lest it make me weak, for I must return to the fighting without delay. Nor is it fitting that I should pour a libation to the gods with my hands uncleansed from the blood and dust of battle. But do you, mother, go to the temple of Athene, together with all the wives of Troy, and make her an offering and pray that she will hold back the Greeks from our city and break the strength of Diomedes, son of Tydeus. While you do this, dear mother, I shall go to Paris and bid him arm himself and come out to fight. All this trouble he has brought on us, and yet, too often, like a coward, he stays at home and lets others fight to keep Helen for him.' And though she was unwilling to let him go, he embraced her and hurried on to the house of Paris, which stood close by the palace.

Having watched him go from her sight with anxious eyes, Hecuba made haste to call all the married women of Troy to the temple of Athene, that they might make an offering to the goddess who, more than any other of the immortals, had given help and counsel to the Greeks since they had sailed for Troy. For the offering Hecuba chose out from amongst her embroidered robes the richest and most beautiful, of many colours and intricate design, which had been brought to Troy from Sidon, far across the sea.

Then she and the other women gathered in the temple, where Theano, wife of old Antenor, Priam's counsellor, was priestess. There, while the women raised their hands to immortal Athene and implored her help, Theano took the lovely robe from Hecuba and laid it across the knees of Athene's statue in the holy place, praying her to hold back the Greeks from Troy and to break the strength of Diomedes.

But Athene loved the Greeks too well to hear the prayers of the women of Troy.

Hector found Paris in his house, busied about his fine armour, polishing here an ornamented breastplate and there a gold-studded baldric or a shield-strap, fitting on his greaves with their silver ankle clasps or trimming the crest of dyed horsehair on a helmet, while Helen sat close by, spinning, in silent discontent.

'Most wretched brother,' Hector exclaimed, 'the Greeks are almost at the gates of Troy, and you disport yourself at home. If you saw another man hold back from battle, you would blame him for a coward who shirks danger. Yet, in this war that is being fought for your sake, little help do you give our people. Put on some of that armour and come out and fight, if you have any pride or courage.'

Paris, with easily assumed penitence, said quickly, 'You rebuke me most justly for my tardiness, good brother, though, truly, it was through grief for the woes that I have brought upon the Trojans, and not through lack of courage, that I lingered here. Indeed, I was even now choosing out my arms to go forth to the fight, and Helen was bidding me make haste.' He smiled beguilingly, very sure of the charm that had so often won him his own way. 'If you will but wait for me, Hector, I will come with you now. Or, if you cannot wait, I will hurry after and over-take you.'

But Hector was no longer to be persuaded by soft words and he looked at his brother in silence, with a contempt that was so familiar that its edge had long been blunted.

Helen flung down the spindle which she held and rose and came to Hector. 'I wish that I had died before I brought so much sorrow on so many people.' Her voice was shrill and biting. 'But if these things are as the gods will them, then I wish that the man for whose sake I left my husband's house had been a better man, or at the least, one who would have paid heed to the censure of others, and felt shame at their scorn.' Then she sighed, and after a moment shrugged her shoulders slightly, smiled at Hector, and said, in a voice more like her own, 'But you look weary, good Hector. Come, sit and rest a little while.'

Hector smiled kindly in return, but shook his head and answered, 'I cannot stay, Helen, for our men need me too badly. Do not try to persuade me, for I must go. Yet, if you can, Helen, put some courage into this brother of mine and bid him hasten after me, that we may go out to the battle together. I shall go first to find Andromache, for things go so ill with us that only the gods know if I shall see her again after today.'

Leaving them, Hector went to his own house, but there he found neither Andromache, his wife, nor his little son, Astyanax. He called to one of the slaves, 'Where is your mistress? Is she gone to the temple of Athene with the queen?'

The woman came forward. 'Our mistress, lord, went in haste to the wall, when she heard that the Greeks had carried the fighting close to the gates.'

Hector turned and hurried from his house, down the

broad street from the Pergamus towards the Scaean Gate; and as he neared the gateway, Andromache saw him and came running to him from the watch-tower beside the gate, followed by the nurse who carried Astyanax.

Hector smiled when he saw the child, but Andromache was weeping. She took hold of his hand. 'Your courage will be the end of you, dear husband. Must you go again to battle? Think of me, if you will not consider yourself. My mother is dead, my father died when the town of Thebe was taken, and my seven brothers Achilles slew. I have no one left but you, Hector. To me you are not only husband, but father and mother and brother as well. If I lose you, what have I left? Have pity on me, Hector, and on our son. Stay with me on the wall today. Call the Trojans together and let them take their stand by the wild fig-tree. For opposite the old fig-tree the wall is weakest, and three times today have I watched the Greeks attack at that place.' She held his hand in both of hers and clasped it close to her.

He put his other arm about her. 'My dearest wife, if I stayed with you and kept from battle, how could I face our people ever again? Indeed, I think that I could not keep from the fighting even if I would; for, all my life, I have learnt only to be valiant and first in danger, to lead other men where they should follow, and to earn honour for my father and myself.' He sighed. 'I think that we cannot win this war, Andromache, and that there must come a day when Troy will fall to the Greeks. My heart is torn when I think of all who will suffer then, my mother, my father, my brothers and sisters, our people; but for none of them do I grieve as I grieve for you, my wife, whom some Greek will take as the spoils of battle. Then, far away over the sea,

in Greece, in the house of a harsh master, someone, some-
day, will see you weeping as you work at the loom or
fetch water from the spring, and he will say, "She was the
wife of Hector, the greatest warrior among the Trojans."
And, hearing, your tears will flow yet faster for the husband who was not able to save you.' He paused and then
broke out, 'Oh, may I be dead and the earth heaped high
above me, before I hear your cries as they carry you off
into slavery.'

Andromache wept; and Hector turned from her and,
leaning forward, stretched out his hands to take Astyanax
from the nurse. But the child was afraid of the great horsehair crest on his father's helmet, and shrank back into his
nurse's arms, so that Hector and Andromache, for all their
grief, had to smile, and even laugh a little, as Hector took
off his helmet and laid it on the ground, and Astyanax,
reassured, came willingly into his arms.

Then Hector kissed his son, and holding him, prayed for
him. 'Great Zeus and all you immortal gods, grant that my
child shall be brave and prove to be, even as I am, the
greatest warrior amongst the Trojans, and grant that he
shall rule mightily, here in Troy. And let there be a day,
when, coming back from battle, it will be said of him,
"He is a far better man than his father," so that he may
gladden his mother's heart.'

He put Astyanax in Andromache's arms, and she smiled
at him through her tears. He held her and the child close to
him, and said, 'Do not torment yourself too much, dearest
wife, for no man can send me down to the land of Hades
before the time allotted by the immortal gods. Now go
home and busy yourself with the household tasks, that you

may forget to fret for me. War is not for women, but for men. And of the men in Troy, for me above all others, who should one day rule here.'

He kissed her, and put on his helmet, while she went up the street towards the Pergamus, turning back time after time to see him, though her eyes were all but blinded by her tears.

Hector hurried on towards the Scaean Gate, and there Paris, running from his house, with all his shining armour on, and laughing as though he had no cares in all the world, came up with him, in a mood more fitted to one who went out to dance than to fight. Hector looked at him with disapproval, but Paris said respectfully, with a sudden change of mood, 'I have kept you waiting. I am sorry.'

For a few moments Hector did not answer, but went on looking at him, yet with an altered expression. Then he said, 'You are not truly a coward, for I have seen you fight well. But you are remiss and slack and careless of your good name. It hurts me and makes me angry when I hear other men speak ill of my brother, and justly.' He smiled a little and laid a hand for a moment on Paris's shoulder. 'But come now, let us go, there is much for us to do. We will make things well again between us, when we have turned this trouble from our city, and the last Greek has left the land of Troy.'

With Paris by his side, Hector made his way back swiftly to the fighting, and the Trojans were glad indeed to see him come, and took heart.

And flashing-eyed Athene came from high Olympus, whilst bright Apollo hastened from his temple on the

Pergamus; the one ardent for the Greeks and the other for the Trojans, and they confronted one another, unsmiling and implacable, beside the sacred oak-tree before the Scaean Gate.

'Why have you come here again, Athene? Is it to give victory to the Greeks?' asked Apollo. 'Must you for-ever afflict the Trojans, you and Hera, jealous of Aphrodite?'

'And must you ever protect the Trojans from a fate which they deserve, and try to ward off ruin from them?' retorted Athene, with anger.

'The outcome of this war is the will of Father Zeus,' replied Apollo. 'Yet for today, let both Greeks and Trojans cease from strife. Let us cause Hector to challenge to single combat some leader of the Greeks, and so may all other men, save those two, fight no more today.'

Athene inclined her head. 'It was with that in mind that I myself came from Olympus. There has been slaughter enough amongst them for one day.'

So immortal Apollo put it into the mind of Helenus to go to Hector's side and say, 'My brother, it nears evening and we are weary. Would it not be a good thing if you were to call upon both Greeks and Trojans to cease from fighting, and challenge whoever is accounted the best amongst them to match himself with you in single com-bat? Thus might we have a respite from fighting, and you may win great honour for yourself.'

For the second time that day, Hector took heed of the advice of Helenus. He bade the Trojans cease from battle and made it known that he would speak both to them and to the Greeks; and as soon as Agamemnon was aware of his intention, he commanded the Greeks, also, to hold back from the fight. The warriors flung down their weapons and sat upon the ground to rest, and Hector stood alone between the two armies and made his challenge. 'Let one of the kings and princes of the Greeks come forward now as their champion against Hector, King Priam's son, to see which is the better man. And let him declare before Zeus and all the immortal gods, that if he chance to slay me, then he will take my armour and my weapons, but leave my body for my people. For my part, I swear that if divine Apollo give me the victory, his armour I shall take and hang up in Apollo's temple, but his body I shall give back to the Greeks, that they may burn it and heap up for him a burial mound upon the shore; so that in days to come, when all we here are no more, some man, sailing over the sea to the land of Troy, may see it and say, "This is the burial mound of a warrior who died long ago, slain by mighty Hector in the olden days." And so shall my fame live on.' Then, having spoken, he was silent, waiting for an answer to his challenge.

Seeing Hector standing there, so strong and eager to do battle, there was no one amongst the Greeks who wished to face him; not even Diomedes, who had done such great deeds that day, for he was wounded and his shoulder gave him pain.

Then, amidst them, Menelaus got to his feet, his tawny beard and hair glinting in the low-lying rays of the late afternoon sun, as he looked all about him. 'Have you become weak women, you men of Greece? Is there no one amongst you with courage enough to meet the son of Priam?' He paused, and when there was no reply, he went

on, 'I am far from being the best warrior here, yet since
no one better has come forward, then I will meet great
Hector and save the honour of the Greeks. The gods will
give the victory as they choose.' And he signed to his
followers to bring his helmet and his weapons.

But instantly Agamemnon sprang up and came to him,
seizing him by the arm and saying urgently, 'Are you out
of your mind, Menelaus? Hector is a far better warrior
than you, and he will surely kill you. You shall not fight
with him, I forbid it. Sit down, and I will find another to
do battle with him in your place.'

At that moment old Nestor rose and stood leaning on
his spear, looking around him frowningly. 'Truly, the men
of today are not what their fathers were. I would that I
were young again, that I might meet Hector and show you
how to fight.'

Shamed by his words, Agamemnon remained standing,
and Diomedes got to his feet, ready, in spite of his wounded
shoulder, to meet Hector if no one else were willing. After
him Ajax, son of Telamon, and Ajax of Locris quickly
rose, and Idomeneus of Crete and Meriones, his friend,
Odysseus, and, too, King Eurypylus, who had come from
snowy-crested Titanus, in mountainous Thessaly, with
forty ships.

Nestor looked at them with approval. 'That is a more
fitting sight,' he said. 'Let us now cast lots so that the immortal gods may decide which one of you shall fight with
Hector.'

They marked their lots and cast them into Agamemnon's helmet, Nestor shook the helmet, and it was the lot
of Ajax, son of Telamon, that was shaken out.

When Ajax knew that the lot had fallen to him, he armed himself saying, 'Now pray to the gods for my victory, my friends; and let Hector beware, for we breed brave men on Salamis.' Then, when he was armed, he took up his huge shield made of seven layers of bulls' hide, covered with a plate of bronze, hanging it about his shoulders, leaving his two hands free to fight with, and strode towards Hector, a tall man with a long shadow, smiling grimly; so that even Hector was troubled at the sight.

And Athene and Apollo, in the likeness of two vultures, sat upon a branch of the sacred oak-tree, to watch the combat with far-seeing eyes.

Ajax stood facing Hector, brandishing his spear, and he cried out, 'Now, Hector, son of Priam, you shall learn that the Greeks have other good warriors besides Achilles. Come, begin the fight.'

Hector raised his spear, and its bronze head reddened in the light of the evening sun. 'Ajax, son of Telamon, I am no woman or untried boy to need your teaching. I have been tested in many a battle, so have a care.' He flung the spear, and with such force, that it pierced the bronze of Ajax's shield and all but passed through to his breast-plate.

At once Ajax cast his own spear, and it passed right through Hector's shield and on through his breastplate, grazing his side. Instantly each of them drew out the spear from his shield and leapt against the other; and this time Hector's stroke failed even to pierce the bronze of the huge shield of Ajax; but the point of Ajax's spear went once again through Hector's breastplate and pricked him low in the neck, so that he gave ground swiftly, but did not cease from the fight. He took a jagged stone and flung it so that Ajax's shield rang loudly with the blow as he held it up before him, and the brazen surface was dented. Thereupon Ajax caught up an even larger stone and flung it with such force that, striking full on Hector's shield, it bore him to the ground. Yet he was up and on his feet again immediately and reaching for his sword, even as Ajax drew his, and they would have fallen upon each other, hacking at shield and helmet, had not the two heralds, Idaeus and Talthybius, come, the one from the Trojan ranks and the other from amongst the Greeks, and parted them with their staves.

'Lords,' said old Idaeus, 'cease your fighting, for you have both proved yourselves mighty spearmen, and now night will soon be on us.'

Lowering his sword, Ajax stepped back, saying to Idaeus, 'Herald, it was Hector who made the challenge. Let him withdraw it, and I will cease from fighting.'

Hector sheathed his sword. 'You are a good fighter, Ajax,' he said, 'and we shall meet again in battle, if the gods so will it. But now the sun has set and it will soon be dark. Let us, therefore, cease our strife. Yet, before we part, let us give each other gifts in token of our mutual respect.'

Ajax willingly agreed, and Hector came forward and gave into his hands a silver-studded sword with a leathern scabbard and baldric; while Ajax took off his richly ornamented crimson belt and offered it to Hector.

And so they parted; and the fiercest day of fighting that

Greek or Trojan had yet known was over, and soon only the dead were left upon the plain, lying all amongst the trampled flowers.

Dusk fell, and from the sacred oak-tree by the Scaean Gate the two vultures were gone; god and goddess, for the moment, reconciled.

V

The Wall

THE Greeks returned to their ships, and Agamemnon feasted all the kings and princes in his own hut, ordering a fine young ox to be killed and roasted, that everyone might eat his fill. There was wine from Lemnos in golden cups, and Agamemnon's slaves standing by with pitchers, ready to refill them the moment they were empty; whilst other slaves went amongst the guests with bread and meat; and a minstrel sang to entertain them. Agamemnon was in good spirits; and for Ajax, who had almost worsted Hector, he had his slaves set aside all the best portions of the meat.

After they had eaten and drunk, old Nestor spoke to them. 'Good King Agamemnon, and all of you, my friends,' he said, 'there is a thing which we have not yet done, which it would be wise to do. In all the long years

since we first came to the land of Troy, we have never yet
built ourselves a wall to protect our ships. We have always
believed ourselves strong enough to guard our ships and
huts with shields and weapons alone—and so, indeed, we
were. But for twelve days now Achilles has not been
fighting at our side, and we feel the lack of his high young
courage and the strength of his Myrmidons, good warriors
all. And today, in the fiercest fighting we have yet seen,
the Trojans sent many of our good comrades down to the
land of Hades. It were well, therefore, that we built a
rampart and a trench before the ships as a defence against
our enemies. Tomorrow, when we have burnt our dead
and raised above them a high burial mound, let us build a
wall to protect our ships and our huts.'

Agamemnon and all the other kings and princes of the
Greeks agreed that his counsel was wise. 'Tomorrow,' said
Agamemnon, 'we shall hold from the fighting and build a
wall. Those are my commands.'

In Troy, meanwhile, on the Pergamus, outside the gates
of the palace, the Trojans held an assembly by torchlight,
and old Antenor spoke before the others. 'Good King
Priam, and all you men of Troy,' he said, 'this war with
the Greeks has gone on long enough, it is time that it was
ended. And so indeed might it have been today, had we not
broken our oath of truce. I say to you, let us give back
Helen, and any treasure she brought with her from Sparta,
to the sons of Atreus, that the one may have his wife again
and the other may sail home to Mycenae, together with
all those whom he called upon to fight against us.'

Having spoken, Antenor sat down again, and there were

many there who murmured in agreement with him; but Paris, his handsome face shadowed with annoyance, rose and replied immediately, and with high-handed insolence, 'Lord Antenor, I do not like your words, and I hope that they were only spoken in an old man's thoughtless moment, and were not intended to move others to the folly of approval. I here declare to you, Antenor, before all the men of Troy, that I will not give up Helen.' He looked about him, and seeing how few there were who approved of his assertion, a little of his self-assurance left him and he went on, almost as though he pleaded with them, 'Whatever treasure she brought with her from Sparta, that shall I restore willingly to Menelaus, and add to it more of my own wealth besides, as much as is fitting—or as much as he demands.' He glanced about him again, then added defiantly, 'But I will not give up Helen.' He sat down, with his face averted, so that he should not see, in the torchlight, the disapproval—and worse—in the eyes of those near enough to him.

For several minutes no one spoke, then slowly and wearily Priam rose. 'Men of Troy, you have heard the answer made by Paris, my son. I, your king, support that answer. At dawn let Idaeus the herald go to the ships of the Greeks with the offer made by Paris. And, further, let him ask if they are minded to keep from fighting until both they and we have gathered our dead from the battlefield and built a burial mound for them.' So Priam spoke, though, he, too, longed for the war to be over; but Paris was his son, and it had never been his way not to stand by his own. And because Priam was their king, the Trojans agreed with his words, and the assembly was broken up and each man went to his home for the night, save those whose duty it was to guard the walls.

At the first light of morning Idaeus the herald went to the Greeks and spoke to them, telling them as Priam had bidden him.

When he had done, they sat in silence, and most of them were thinking how they were weary of the war, and there was more than one man there who reflected that, when all was said and done, Helen and her misdeeds were no concern of his, since she was not his wife, but the wife of another man. And everyone was very careful not to look at Menelaus.

But Diomedes was one of the younger kings amongst them, more ready than most for battles and fighting, and, moreover, he had not yet lived long enough to have had to learn to compromise. He then rose and said, 'Are we now to take gifts from Paris and be grateful? Paris need give us no treasure; no, nor need he give us Helen. The Trojans cannot resist us much longer, even a man of little wit could tell us that. If we fight resolutely and with determination, Troy must fall; and then we can take both Helen and all the treasure we want, a hundredfold what Paris offers us.'

And because there was no one there who would willingly have seemed to be of less courage than Diomedes, they all agreed with him, and counselled that the offer made by Paris should be rejected.

Agamemnon rose. 'Herald, you have heard the answer of the kings and princes of the Greeks to the offer of the son of Priam. My own answer is even as theirs. But concerning the dead: let both Greeks and Trojans keep truce on this one day, that their dead may be burnt fittingly.' And in the name of the immortal gods, he swore to keep from fighting for that day.

Before the sun was high, both Trojans and Greeks had gathered their dead and built for them two great pyres, the one near the gates of Troy and the other close by the ships; and when the flames had sunk above the ashes of the pyres, they raised burial mounds above them, that all those whose bones lay beneath might be remembered by their comrades.

Then, for the remainder of that day, as Agamemnon had ordered, the Greeks built a wall just inland of the first line of ships, to be a protection to their camp. First they dug a deep trench and planted it within with sharpened stakes. The earth from the trench, together with stones and rocks and baulks of timber, they built up into a rampart, to the seaward side of the trench; and they worked with such a will, that by nightfall their wall stretched from before the ships of Ajax, at the most eastern point, to the ships of Achilles, at the very west. Here and there in the wall they had set narrow gates, strongly barred, and, at the mid-most point, they had made a wider gateway and a track for their chariots to pass out on to the plain.

Early the next morning, battle was joined once more, and again the plain resounded with the clash of weapons and the shouts and cries of men, and the shrill neighing of the horses as the chariots sped here and there.

But, on Olympus, Zeus called all the gods and goddesses before him and forbade them, on pain of his most fearful wrath, to join in the battle and to fight for either Greek or Trojan.

However unwilling certain amongst them might be, they were silent at his words and bowed their heads in obedience to the command of the king and father of them all, and only Athene dared to stand before him and speak.

'Father Zeus,' she said, 'we know your might and we dread your wrath, yet there are those amongst us here in high Olympus who pity the Greeks, the unhappy Greeks, who fight and perish far from their own homes. We shall hold from battle, as you command us, Father Zeus, yet for their comfort and to their profit, those of us who love the Greeks will offer them our counsel.'

So she spoke, openly and unabashed before him, as was her wont, and Zeus smiled, for he loved her above all his daughters. 'Do not despair for your Greeks, my child,' he said, 'for, in the end, things will be well for them.'

Then, in his chariot drawn by his two immortal steeds, whose hooves were flashing bronze and whose gleaming manes were thin threads of gold, as dazzling as the sunlight, he came from Olympus to Mount Ida, within sight of Troy—Mount Ida of the many springs, where the wild beasts roamed; and there, upon the topmost peak, he sat to watch the fighting of Greeks and Trojans.

Until midday the battle was equal, with neither Greeks nor Trojans gaining ground, nor yet giving way, one before the other. But in the afternoon, in fulfilment of his promise made to Thetis, Zeus filled the Trojans with might and courage, so that they hurled themselves upon the Greeks, who were driven back towards the shore.

One by one, the kings and princes of the Greeks, who

had been foremost in the fighting, turned their chariots and made for safety before the Trojan onslaught. Back towards the ships went Agamemnon and Idomeneus, Ajax, son of Telamon, and Ajax of Locris.

But when old Nestor would have turned his chariot, Paris, from a distance, killed one of his horses with an arrow, terrifying the other, which plunged and reared, so that the chariot was almost overturned, and Nestor's charioteer, thrown into confusion, could make no move to help his lord.

Nestor climbed down from the chariot, sword in hand, to cut the dead horse free; while Hector, from a short way off, seeing him to be at a disadvantage, made towards him, coveting his shield of solid gold, for it would have been reckoned a fine prize in Troy.

Diomedes and Sthenelus, however, saw the old king's plight, and hastened to him, Sthenelus urging on the horses of Aeneas, which he drove. Seeing Odysseus turning for the shore, Diomedes called out to him to come with them to the aid of Nestor, but Odysseus did not hear him and went on, and Diomedes and Sthenelus came to Nestor alone.

'Good old king,' said Diomedes, 'with one of your horses dead, and a worthless weakling for your charioteer, you can never hope to escape death at Hector's hands if you wait here for him alone. Come, instead, into my chariot and see how fleet these horses are that I took from Aeneas two days ago. Well have they served me today. Take the reins from Sthenelus and mount beside me in his place, and he shall take your chariot and your horse safely from the battle to the ships, while we two meet with Hector.'

Sthenelus leapt down and, gladly, Nestor mounted beside Diomedes. He took up the reins, and together they drove to meet Hector's attack. Diomedes flung his spear, but missed Hector, and struck, instead, his charioteer, who fell wounded. Instantly Hector checked his horses, and turned them aside, calling to his followers, and at once a man ran forward to take the place of the fallen charioteer; and Hector once more faced Diomedes' attack.

But that victory that day might be with Hector and the Trojans, Zeus hurled down a thunderbolt before the chariot of Diomedes, and the horses of Aeneas were terrified, and would not move forward a single step, but stood trembling, with wild eyes.

'It was a warning from immortal Zeus,' said Nestor, shaken, 'we would do well to heed it. Let us leave pursuing Hector and return at once to the ships. For no man, however valiant, can prevail against the will of Zeus.'

'That is true, good Nestor,' Diomedes replied. 'Yet I would not have Hector say of me, before the Trojans, that the son of Tydeus fled from him. I would rather the earth opened to swallow me, than that Hector should make such a boast.'

But Nestor rebuked him sternly. 'That is a foolish thing which only a young man would think to say, Diomedes. Let Hector boast in Troy, if he will. And if he calls you coward, will the Trojans believe him, who have seen you in battle? Or will the widows believe him, whose husbands you have slain? No, Diomedes,' he went on urgently, 'it is ill to ignore a warning from the gods. I shall not permit such folly.' And with that he turned the horses, who were only too willing, and drove towards the shore.

When Hector saw them fleeing he was glad, and bade his charioteer make haste after them, calling himself to his horses, 'Come, my good horses, pay me now for your keep. Remember how often my dear Andromache has herself set corn before you. Many times has she watered you when you came thirsty from the battlefield—often before she poured wine for her own husband. Come now, friends, for old Nestor's golden shield would be fine spoils to take home to Troy.'

After Hector, across the plain, raced the other chariots of the Trojans, and, not far behind them, the warriors who fought on foot. And the Greeks could not withstand their attack, but retreated to the wall that they had built and poured in through the gateway and closed it fast; so that by the time that Hector reached the trench, there were none but dead or wounded left outside.

Agamemnon came down from his chariot, flinging off his purple cloak, and hurried to the assembly place, and standing there, he lifted up his voice, chiding the Greeks who had fled from battle, and bidding them prove their courage and go forth to attack before night fell. Those who heard his words took heart from them, and passed them on to the others; and, while Hector drove his chariot up and down before the trench, vowing he would break down their wall and fire their ships, they rallied and gathered before the gateway to attack.

Then, on a signal, the gates were flung open and the Greeks came forth, the chariots first and the men who fought on foot close after them. First went Diomedes and Sthenelus, with the horses of Aeneas, and well they carried their new masters, swerving here and there amongst the Trojans while Diomedes struck down man after man.

Hard behind Diomedes came the sons of Atreus, Menelaus and Agamemnon himself, then Ajax of Locris, Idomeneus and Meriones, Eurypylus of Thessaly, and Ajax, son of Telamon, with Teucer, his half-brother, a famed archer.

These two sons of Telamon soon leapt down from their chariot and fought side by side, as was often their custom, tall Ajax holding his huge shield to shelter Teucer as he bent his bow, peering out from behind the shield like a tortoise from its shell. Many Trojans fell to his arrows; but it was Hector whom he longed to kill. Twice Teucer aimed at Hector, and twice Apollo turned the arrow aside. Then Hector leapt down from his chariot, and taking up a large stone, he hurled it at Teucer just as he had fitted yet another arrow to his bow and was drawing back the string. The stone struck him on the neck with such force that he fell senseless and the bowstring snapped. But, though Hector leapt forward with a shout of triumph, he could not reach him, for Ajax stood to guard his brother until two of their comrades had borne him back within the wall to safety.

Then once again Zeus put great courage into the hearts of the Trojans, and great strength into their hands, and once again the Greeks fell back before their onslaught and retired within the wall, drawing the gates close after them with difficulty and leaving many dead or dying in the trench or on the plain before the wall.

Within the wall they lifted up their hands and prayed to all the gods for help; but, outside the wall, Hector drove his chariot here and there, rejoicing in his victory and ever threatening to burn the ships:

But it was evening, and fast growing dark; the sun had gone and all battle was over for the day. And never since they had come to Troy had a night been more welcome to the Greeks; though the Trojans would not have had it come so soon.

On Hector's command they withdrew, beyond the battlefield and the slain, to the banks of the River Scamander, and there they made their camp for the night, unharnessing the horses from the chariots and building fires, while some hastened back to the city to fetch food and wine, and fodder for the horses.

Hector stood amongst them and spoke to them, exultingly, 'Men of Troy and brave allies, the night has thwarted us, or we should have made an end of the Greeks this very day. Yet let us eat and drink and rest tonight, and see what the dawn will bring to us. For I do not doubt that tomorrow will be even as today, bringing great glory to us all, and disaster to our enemies. And I think that tomorrow the son of Tydeus shall fall to my spear, for the immortal gods are with us, my friends, and they have abandoned the Greeks. Yet let us not forget to keep our fires burning brightly until the dawn, and hold good watch; not for fear that the Greeks attack us in the darkness, but lest they seek to sail away unseen. They shall not easily leave the shores of Troy, who came here uninvited.'

All the Trojans shouted aloud in triumph at his words, beside themselves with the day's success.

But Hector was not satisfied to chance the Greeks' escaping him. After the Trojans had eaten and drunk their fill of the good things brought from the city, he called to him the lords of Troy and the most trusted amongst his men, asking for someone of them who would go alone to the camp of the Greeks and spy out, if he could, their intentions: whether they were preparing to set sail, or whether they would stay and fight again in the morning; and to learn whether, in their despair at the way things had gone against them, they had become careless about keeping guard, or whether, in great fear, they dared not even sleep, but sat, every man, watchful and alert, afraid of a surprise attack by dark. 'To the man who ventures this deed,' said Hector, 'I will give the finest chariot and the two best horses which we take from the Greeks as spoils. Now, who out of you all will do this thing for me?'

At this a man named Dolon rose, an ill-favoured man, but a good runner. 'I will go to the ships of the Greeks for you, Lord Hector, and spy out their counsels,' he said, 'if you will swear to me that, as my reward, you will give to me the chariot and the horses of Achilles, son of Peleus, which, if all that I have heard of them is true, are immortal and unsurpassed.'

So Hector swore, by all the gods, that, when the war was over, Dolon should have for himself the chariot and the horses of Achilles. And later, when night hung dark above the plain, Dolon made ready to steal away, down towards the shore.

VI

The Embassy

BUT while the Trojans, flushed with the triumphs of the day, waited confidently for the morning, when they might resume the slaughter, Agamemnon, anxious-eyed and sick at heart with his misgivings, sent his heralds to call the kings and princes of the Greeks, and their lords, to the assembly place upon the beach.

When they—as apprehensive as their leader—were all gathered in their places, he stood up before them and spoke. 'My friends, when we set sail for Troy, it was with good omens, and I believed that immortal Zeus would give the victory to us. But Zeus has cheated me, and now he plans to give the victory to our enemies, and we shall all perish here, before the walls of Troy, if we stay longer in this place. So let us rather, while it is yet not too late,

70

leave the cursed shores of Troy and sail for our homes in Greece.' He finished, and all the Greeks were silent at his words, while he looked from one man to another of those gathered there, scanning in concern each man's face in the fading light, while he waited for their reply.

But no one spoke, until at last Diomedes stood up. 'It was not so very long ago, King Agamemnon, that you taunted me with cowardice, and I did not answer you. A strange taunt from one who now counsels us to fly from Troy! You say that Zeus has cheated you, King Agamemnon. I know nothing of that, but I know this, at least: that, at your birth, Zeus endowed you in unequal proportion. He gave you wider lands and greater riches than any other king, so that you have leadership over us all; but he never gave you the courage which a great king should have. Because we have followed you to Troy and are subject to your commands, do you think that we are all cowards like yourself? If you want to go home to Mycenae, go home. There is the sea, and there are your ships. No man will hinder you. But I do not doubt that there will be found few amongst us to follow you. You will have the whole broad sea to yourself on your return to Greece.' He paused and looked about him at the others, hoping for their approval of his words, but they still sat silent. Yet, though no one spoke in agreement with him, nor did anyone speak for Agamemnon, and Diomedes waited.

Then, after a while, his head held high and his eyes flashing, his voice rang out again across the assembly place. 'Go home, all of you, with him, if you will. But when you are all fled, yet will there still be two Greeks left on Trojan soil, for Sthenelus and I fight on, to the very walls of Troy, until Priam's city falls to us alone.'

As he ceased, a great cheer rose from all the Greeks, and they shouted aloud for him, praising his courage, their former dismay now quite forgotten, while Agamemnon frowned, torn by indecision.

When the cheering had died down, Nestor rose, and he was smiling. 'It does my old heart good to hear a young man speak with courage, as you have done, Diomedes. You are brave in battle, and though you are amongst the youngest of the leaders of the Greeks, yet your words need be despised by none. No, not even by those older and more powerful than you. But now, my friends, all you kings and princes of the Greeks, I would offer you my counsel. I am an old man. I have fought in more battles than you have seen, and had as comrades men finer than any you have ever known, so do not scorn my advice. What I have to say concerns you all, but it is to King Agamemnon that I would address my words. And because I am old, and the old are privileged to speak their minds, do not be angry, great king, if what I say displeases you. For you are the mightiest amongst us and the leader of us all, and therefore it is your duty, not only to give the best commands, but to listen to the wisest counsel, also. Rashly, and unadvised save by your own folly, and against my words, who sought to counsel you to peace, did you take from Achilles his lawful prize to satisfy a grudge, and he withdrew his help from our most righteous cause. Since that day we have lacked his help against the Trojans, and bitterly have we regretted it, for not only have his courage and his fighting skill deterred our enemies in the past, but his

Myrmidons are all brave, well-tried warriors and we feel their loss most keenly. All the troubles that have lately fallen on us come from this one cause: that you took from another that which was his by right, and put dishonour on a proud and headstrong youth. It was ill done, good king, and it were best you made amends, and swiftly. Send back the woman Briseïs to Achilles and give him gifts, that he may know you are his friend, and may once more bring his Myrmidons into the field against the Trojans, and so give our men new heart and turn the tide of battle for us.'

All were silent again when he had spoken, while Agamemnon's dark thoughts twisted and turned in his mind as he pondered what he should do. But because he was the leader of them all, and their safety rested in his hands, he said at last, bitterly, 'Fool that I was to forget that one man whom great Zeus loves well and honours is worth more to him than a whole army. Zeus and the immortal gods must love young Achilles well—for is his mother not herself a goddess?—and for his sake will see all our mighty army utterly destroyed. No king, however great, can hope to prevail against one so well loved by the immortal gods. For the sake of you all, that you may not perish here, and for the sake of my good brother Menelaus, that he may not be compelled to return to Sparta without the wife who was stolen from him, I will make peace with Achilles. I will return the girl Briseïs to him, and give him goodly gifts besides, and so shall he be won to fight again for us. I will give him gold, and shining vessels, and seven new-gleaming tripods which have never yet been set above the fire; and twelve of my finest horses shall be his. From the spoils we took from the island of

Lesbos, which he himself won for me, I will give to him those seven women, well skilled in weaving and all women's crafts, whom I chose for my share. They shall go to him with Briseïs. All these things will I give him now, from the riches that I have here with me; but when Troy is taken, he shall have more besides. He may make first choice of the captives, even before I myself have chosen; and when I am back home in Mycenae, he shall always be welcome in my palace, and honoured there as though he were my son. And more, he has no wife, and I have three fair daughters. He shall take his pick of them and carry her home to Phthia and pay me no bride-price, but I will, for my part, give her a greater dowry than ever any king's daughter had: seven towns, close by the sea, where the pastureland is good and the harvests rich. All this shall I do for him if he will put away his arrogance and spleen and submit to my authority as well he should, for am I not his elder? Aye, and a far greater king.'

'That is well spoken and generously offered, King Agamemnon. Surely not young Achilles nor any other man could refuse such noble gifts as you have named. Come,' said Nestor eagerly, 'let us send to him at once and tell him of your offer, and bid him forget his anger. Let Odysseus, who speaks well and is skilled at moving men's hearts, go to him, and Ajax, son of Telamon, and your two heralds, King Agamemnon. And, above all, let good Phoenix go with them, for he was Achilles' tutor, and Achilles loves him well and will listen to his words.'

And so, when they had poured libations to the gods and drunk, and prayed that Zeus might have pity on them and turn Achilles' heart, Odysseus and Ajax set out, accompanied by the two heralds, with old Phoenix hurrying on a little way before them, to prepare Achilles for their coming. They went along the shore in the evening light, with the sounding of the restless waves upon their right hand, and they were silent as they went, each man praying in his heart that Achilles might be easily persuaded.

When they came to Achilles' ships, they passed through the stout gate, bolted with a fir trunk, which was set in the palisade of stakes that surrounded the huts of Achilles and his men. In the wide enclosure they found the Myrmidons preparing their evening meal, building fires in the open space and spitting huge joints to roast above them.

Entering the pillared outer porch of Achilles' hut, to go into the hut itself, they heard from within the sound of a clear voice singing, and found old Phoenix, a smile upon his lips, waiting in the doorway as though he were unwilling to interrupt the singer. Still smiling, he held up his hand for silence as the others came near, and made no move to enter; but Odysseus looked over the old man's shoulder to see who it was who sang so sweetly. And there he saw Achilles with Patroclus, just the two of them, alone, save for their hounds lying by the hearth; and it was Achilles who was playing on a lyre and singing to his friend of the deeds of the old heroes.

Odysseus, noting with satisfaction that Achilles seemed to be in the best of moods, thought that this promised well for the success of their mission, and, despite the old man's protests, set Phoenix gently aside, and went into the hut.

Achilles, glancing up to see which of his men it was had entered, saw, with astonishment, Odysseus with Ajax close after him, sprang to his feet, and, still holding his lyre, went forward to welcome them with unfeigned pleasure. Patroclus turned at the interruption, and then rose also, with equal surprise and gladness. For it seemed long to both of them since they had spoken with any of their former battle-comrades.

Achilles greeted his guests kindly. 'Though you stand by Agamemnon in his quarrel with me, yet are you still my good friends, and very welcome.' He embraced his old tutor and led the three of them to seats spread with purple-dyed rugs, calling cheerfully to Patroclus, 'Fetch a larger bowl for the wine and let us mix less water with it, now that we have companions to drink with us.'

After they had poured libations to the gods and drunk, Achilles said eagerly, 'You are in time to share our supper. Will you stay and eat with us?' And he called for meat to be brought. Automedon, his charioteer, came in with a side of mutton and the chine of a fat pig and laid them on the cutting block which Patroclus had set ready, and Achilles cut a joint from each while Patroclus blew up the fire. Then the meat was sliced and spitted over the hot embers; and when it was cooked, Achilles laid it on platters for his guests, while Patroclus handed round bread in a basket. Then Patroclus cast into the fire the portion which was the offering to the gods, and they all sat down to eat.

When they had eaten and drunk as much as they would, and Achilles was in good spirits, Ajax looked across at old Phoenix, eyebrows raised in query; Phoenix nodded, and Odysseus, watching, knew that the moment was favourable for him to speak. He filled his cup with wine and pledged Achilles, saying, 'The best of welcomes have you

given us, and a fine supper, Achilles, and we thank you for
it. Yet are our minds not truly set on meat and drink, and
our hearts not wholly given up to the good company, for
we cannot forget the peril in which the Greeks now
stand, and our hearts are heavy for our plight. For the
Trojans, led by Hector, whom no man can now withstand,
have driven us back to the ships, Achilles, and we lie tonight between their camp-fires and the sea. And Hector,
impatient for the dawn, has sworn that tomorrow he will
burn our ships. By this time tomorrow, the war may be
over and the Trojans victorious, and we, whom you call
your friends, may all lie dead, unless you will forget your
anger with King Agamemnon and fight with us once
more. Good Achilles, when your father, King Peleus, sent
you with me to Agamemnon to fight against the Trojans,
on that day when he bade you farewell in Phthia, did he
not warn you to keep a curb upon your rash temper, and to
refrain from quarrelling that you might win more respect
from men? I know he told you this, but it seems that you
have now forgotten his wise counsel.' He leant forward
and spoke earnestly, with carefully chosen words. 'But it
is not too late, Achilles. Set aside your anger and forgive
Agamemnon's folly. He has admitted he did wrongly, and
will send Briseïs back to you, and give you many gifts
besides. Seven tripods, twenty cauldrons, and much gold;
twelve horses which have won many races, and seven slave
women, skilled in all handicrafts. These things will be
yours tomorrow; and when Troy has fallen, thanks to
your aid, then you shall have the pick of all the Trojan
captives and first choice from Priam's treasure house; and
evermore in Mycenae you shall be to Agamemnon as a

son. You may choose a wife from amongst his daughters and pay no bride-price for her, and she will bring you seven towns in Argolis as a dowry. All this if you forgive him and bring your Myrmidons to battle in the morning.'

Odysseus paused, and watching closely in the firelight, he saw that Achilles was not to be won by gifts, and that Agamemnon could not buy either his forgiveness or his help, and he said quickly, before Achilles could answer him, 'Or if you cannot find it in your heart to forgive King Agamemnon, then have pity on the rest of us. Do not let the Greeks perish, but fight for them and win their everlasting love and gratitude, and great glory for yourself. For, who knows, Achilles, you might meet Hector himself tomorrow in the battle. He would be a worthy adversary for even you; and the man who kills great Hector will be remembered for ever with honour in the minstrels' songs.'

Achilles looked at him. 'You speak skilfully, as ever, Odysseus. Surely there is not one amongst the Greeks like you for fair speeches and well-chosen words. But even your urging shall not move me.' He made an impatient gesture. 'Come, I have had enough of blandishment and exhortation, here is my straight answer: I will neither forgive Agamemnon nor fight for him. I want none of his gifts. As for Briseïs, let him keep her. She is his now, I do not want her back. Though she was but a woman I won by my spear, she was nobly born, and I would have married her, perhaps. But Agamemnon took her from me. We all came here to fight in this war for the sake of a man whose wife had been stolen from him. Are the sons of Atreus the only men who have the right to cherish their own womenfolk? No, Odysseus. Agamemnon can win back his brother's wife without my help. Or he can perish here on Trojan land. I care not. Go back to him, Odysseus, and you, too, Ajax, and tell him what I say: that I shall never be bought with promises or lured with gifts—not though he offer me twenty times what he has offered, not though his gifts outnumber the grains of sand that lie along the shores of Troy. And as for his daughter, were she as fair as golden Aphrodite and as skilled in women's handicrafts as divine Athene herself, I would have none of her. Let our lord, the great King Agamemnon, who is the mightiest of us all, find a more fitting match for his daughter than Achilles, who will one day rule over no more than little Phthia—if he live long enough. I am doomed to die young, and I have accepted my doom, but my life is too precious to me for me to throw it away for Agamemnon's sake. No, more than that: I would rather lose the glory and un-dying fame which I was promised with my short life, and endure instead the alternative long days of obscurity, with no man to remember me when I am dead, than that my everlasting fame should be won in dying for Agamem-non. I would sooner set sail tomorrow for Phthia and leave him to perish here. And if the rest of you take my counsel, you will do likewise. Let the sons of Atreus finish their own war alone.' He rose. 'Now, go, my two good friends, and tell Agamemnon what I have said. But you, my dear old Phoenix, you must stay here with me tonight, and then, if I do indeed sail home, you must choose whether you will come with me or remain with Agamemnon.'

There were tears in the old man's eyes, but he answered promptly, 'My child, how could I stay here without you?

The gods never gave me a son of my own, but I have always thought of you as a son. When you were very little, you would not go to a meal in your father's hall unless I were there with you, to set you on my knee and cut your meat for you and hold the cup to your lips. And when your father sent you, still no more than a boy, to lead his Myrmidons to fight at Troy, knowing nothing of cruel war or of the assemblies and councils of grown men, he bade me go with you to guide and help you. No, my child, I cannot stay here without you. If you go home to Phthia, I go with you. But I implore you, Achilles, have pity on the Greeks and help them in their despair. There would be no shame in yielding to their pleas. For not even the gods are unbending, even they will answer prayers, and by sacrifice and supplication can a man who has sinned avert the wrath of the gods. For prayers are the daughters of immortal Zeus. Lame and wrinkled and with downcast eyes, they follow hard on the heels of sin. Sin is strong and swift and runs fast ahead, harming and working ill; but the prayers follow after, stumbling and limping, yet never giving up the chase, and healing all hurts they find. Therefore one should show reverence to these daughters of Zeus when they draw near, and deny them not; and thus, respect the gods. My child, let your heart be moved by Agamemnon's prayers, for he has acknowledged his fault.'

But even by old Phoenix, whom he loved, would Achilles not be persuaded. 'My dear old friend,' he said, 'you wrong us both by your tears and by your pleas for Agamemnon's cause. You should rather hate the man I hate. Come now, it grows late and you must sleep.

Patroclus shall spread a couch for you here, close by the hearth.'

Odysseus still sat undecided whether he should again try to prevail upon Achilles, but tall Ajax rose, saying, 'Come, Odysseus, for we can achieve nothing by staying longer. Achilles will not hear us. He has no pity for his comrades.' He turned to Achilles and said bitterly, 'I have known men accept gold in recompense for the slaying of a brother or a son, and lose no honour by it. But you, for one slave girl, you refuse all a great king's gifts.'

'It is not a question of whether I pity the plight of my comrades,' said Achilles sharply. 'It is not whether I would or would not help the Greeks. It is a question of the wrong that has been done to me. I cannot forget and I will not forgive the slight which Agamemnon put on me. Now go, before I am angered with you, who are my friends.'

And after they had poured a last libation to the gods, Ajax and Odysseus went, heavy at heart, to tell Agamemnon of Achilles' answer, whilst old Phoenix lay down beside the hearth to sleep; and, after bidding him rest well, Achilles and Patroclus went to the inner room of the hut, where they slept.

In the spacious hut of Agamemnon, Odysseus and Ajax found all the kings and princes of the Greeks awaiting their return.

'What answer did he give you?' asked Agamemnon eagerly. 'Will he save the ships from burning?'

'Most noble Agamemnon,' replied Odysseus, 'Achilles will have none of you or your gifts, and he bade us tell you so.'

Agamemnon's face fell, as he saw his last hope gone; and everyone there was silent, wondering what would become of them and whether they were fated, all of them, to perish in the land of Troy.

At last Diomedes spoke. 'King Agamemnon, it would have been best had you not humbled yourself to Achilles or offered him gifts. He has ever been proud, and now you have made him prouder still. But let him do as he will, he and all his men, and let us put him from our minds and think instead thoughts of valour: how well we may acquit ourselves tomorrow. There is no profit in sitting here sadly and regretting what might have been, so let us now go and take our rest, and thereby gain what strength we may to serve us in good stead when day is come.'

Agamemnon sought to hold back no man; and so they went, each to his own hut, and lay down to rest and sleep.

VII

The Spy

YET amongst the kings and princes of the Greeks there were two who could not rest that night: the two sons of Atreus lay wakeful in their huts.

Menelaus could not sleep through concern for his friends and comrades, who now faced defeat and ruin for his sake, because of Helen, his wife; and at last he rose, meaning to go and wake his brother Agamemnon, to see if he could gain any comfort from him.

But Agamemnon tossed and turned upon his bed, trying in vain to devise some way out of the plight to which he had brought all the Greeks through his vindictiveness and folly; then, thinking that if any man could advise him, wise old Nestor could, he rose and put on his tunic and sandals, flinging about his shoulders a fine lion-skin, and taking up a spear, he went out from his hut even as Menelaus came to him.

'Why are you stirring, brother?' asked Menelaus. 'Is it to call us to council once again? Have you some plan whereby we may avert our fate?'

'No, Menelaus. I can see no way at all to avoid defeat. Truly, the immortal gods are with Hector and have deserted me. Yet let us all take counsel once again, and maybe someone of us—wise old Nestor or wily Odysseus, perchance—can devise a means to save us. So go, Menelaus, rouse Ajax, son of Telamon, and Idomeneus, and the other leaders whose ships lie farthest from here, whilst I go to good old Nestor.'

Willingly Menelaus hurried away to carry out his brother's commands, glad of any action which would empty his mind of its tormenting thoughts.

Nestor was lying asleep on his bed in his hut, with his armour close at hand beside him, and he awoke at once when Agamemnon came to him, carrying a torch lighted at one of the watch-fires. 'Is it Agamemnon? What is amiss?' he asked, sitting up.

'Good Nestor, I cannot sleep. I fear the dawn; I fear an attack by night. Comfort me with your wisdom. Let us go together and see that the watchmen on the rampart are not sleeping, and let us call our friends together for counsel.'

Nestor rose instantly and pulled on his tunic. 'Let us do as you say, good King Agamemnon.' He sat to fasten his sandals, and paused to frown up at Agamemnon in the torchlight. 'I shall chide Menelaus when I see him, that he has left you to rouse us all yourself. He is at fault to lie asleep. He should be with you now when you have need of him. I know your love for him, great king, yet you must not mind my saying so.'

Agamemnon smiled. 'This time you do him wrong. I know that he is easy-going; and often he seems slow to assert himself, but that is through respect for me, that I may first make known my opinion and my wishes. Yet tonight he rose even before I did, and has now gone to call those leaders whose ships lie farthest from ours.'

Nestor wrapped around himself his thick cloak of shaggy, purple-dyed wool, and together he and Agamemnon went to arouse Odysseus, who came from his hut to them, alert in an instant. But, going next to Diomedes, Nestor found him, in all his armour, lying outside his hut on an ox-hide, his head resting on a rolled-up woollen rug, with Sthenelus close by, and all about them their chosen comrades, in their armour, their heads pillowed on their shields, and beside each man stood his spear ready to hand, thrust into the earth: for Diomedes was taking no chances of a surprise attack. He sat up when Nestor woke him, and when he heard that he was summoned to an assembly, he said, 'Must you go all along the shore, good Nestor, calling the leaders to council? Is there no one younger to run errands for King Agamemnon?'

'There is indeed one younger,' said Nestor. 'You can go for me, if you will, and rouse Ajax of Locris.' And Diomedes went at once, snatching up his spear and running off into the darkness.

Nestor praised all the watchmen whom he found alert at their posts, knowing full well, in his wisdom, that just approbation and well-earned approval can do much to bring courage and confidence to a man.

When all the kings and princes of the Greeks were gathered together, they sat, and Nestor was the first to speak, for it was to him that they all looked for counsel. 'There is nothing,' he said, 'so disheartening as uncertainty. It would be well if we sent forth some brave man to the Trojan camp to spy out, if he can, whether they are stirring and mean to attack tonight, or whether they will sleep and leave all fighting until the dawn. Is there one amongst us who would do this thing and win great honour from his comrades?'

Diomedes said, 'I will go. I am not afraid. But if someone else were to come with me, our advantage would be doubled. For four eyes notice more than two, and what one man misses, another often sees.'

Immediately Ajax, son of Telamon, and Ajax of Locris offered to go with him. Odysseus offered, and Menelaus; Antilochus, Nestor's son, and Meriones the Cretan, the friend of Idomeneus.

Before it could be decided upon who should go with Diomedes, Agamemnon said quickly, 'Good Diomedes, choose which of them you will to be your companion in your dangerous task. Take with you that man whom you know to be most fitted. Do not think that from respect to a man's rank you must choose him to be your comrade and leave a better man behind, because he is less highly born.' So he spoke, because he feared that Diomedes might choose Menelaus, and he did not want his brother to run so great a risk of being taken or slain by the Trojans.

But Diomedes said, 'If I may indeed choose as I wish, then let Odysseus come with me, for he thinks quickly and can get us quickly out of danger. And he is, besides, much loved by immortal Athene. If Odysseus comes with me, I do not doubt that, if we had to, we two would pass safely through even fire and flames, thanks to his cunning and to divine Athene.'

Forthwith the two of them made ready to go. They did not need armour and shining helmets, for the metal of a breastplate would have rung upon the stones if its wearer had dropped to the ground to hide, while bright helmets would have caught the light of the Trojan watch-fires. Instead, Diomedes put on an ox-hide cap, such as was worn by the common fighting men, and by the slingers and archers of Locris, and, since he had left his own sword at his hut, taking up only his spear when he ran to call Ajax at Nestor's bidding, he borrowed a sword to take with him. Meriones lent to Odysseus a curious cap of leather, plated with slivers of boar's tusk, which had been a much-prized guest-gift to Prince Molus, Meriones' father; and gave him, moreover, his own bow and quiver, which he had with him.

Bidding the others farewell, the two of them set out, and, as they left the rampart and the trench, they heard on their right hand the cry of a heron. 'An omen from Athene,' whispered Odysseus, much pleased; and he prayed to her. 'Divine daughter of Zeus, as you have ever been with me in the past, so, I beseech you, be with me tonight. Grant that we may go amongst the Trojans to their cost, and return safely to the ships.'

And Diomedes, too, prayed to Athene. 'Great daughter of immortal Zeus, stand by my side and guard me well; and when I reach the ships again, I will sacrifice to you a heifer with gilded horns.'

So they prayed to Athene as they moved across the plain, amidst the fallen arms and the dead; and Athene heard and granted their prayers.

Cautiously they made their way towards the Trojan camp, under cover of the rocks and bushes which stood here and there upon the plain. When they were about half-way between the ships and the blazing watch-fires of the Trojans, Odysseus caught sight of a quick and furtive shadow coming towards them. He grasped at Diomedes' arm. 'See there, someone is coming from the camp. It may be no more than someone out to rob the dead, but it may be someone sent to spy upon the ships. Let him pass us by, and when he has gone a little distance, let us turn and follow him. In that way he will be between us and the shore, and cannot escape us back towards the city.'

Thereupon Odysseus and Diomedes lay down amongst the slain, as though they, too, were dead; and after a little, suspecting nothing, Dolon—for it was he—went swiftly past them.

When he was a short way on, they rose and ran after him, towards the ships, and hearing the sound of their steps, Dolon stood and waited, supposing them to be his comrades from the camp, come after him to tell him of a change in Hector's plans. But when they were within a spear's throw of him, in spite of the darkness, he knew them for strangers and fled—straight for the ships, since there was no other direction in which he might go.

Rapidly they gained on him, and Diomedes called out, 'Stand, or I shall kill you!' and with that, flung his spear so that it passed over Dolon's head and stuck in the ground a little way before his feet, missing him on purpose; since they did not want to kill him until he had told them all they wished to know.

Terrified, Dolon stood still, and Odysseus and Diomedes came up with him and seized him by the arms. Instantly he cried out, 'Take me alive, for my father is a rich man, and will pay you a good ransom for his son.'

'Who talks of death?' said Odysseus chaffingly, with seeming friendliness. 'Put such thoughts out of your mind. But come, tell us what you do here, so close to the ships in the darkness of the night?'

Trembling, Dolon said, 'Fool that I was, I let myself be lured by Prince Hector's promises. He swore to give me the horses of Achilles, son of Peleus, if I would go amongst the ships of the Greeks and find out whether they had thoughts of flight, or whether they were determined to fight again tomorrow.'

Odysseus chuckled. 'Truly, my friend, you set your hopes high. The horses of Achilles! Indeed, you would not find it easy, I think, to drive the immortal steeds of King Peleus, which the gods gave to him. But tell me, how is the Trojan camp disposed? What watches do they keep? Where does Hector lie tonight?'

'Hector is not now in the camp. He has gone to take counsel of the great lords of Troy at the burial mound of Ilus, nearer to the city. On the camp there is no special guard, save what is customary, around the fires. The allies are all sleeping, they keep no guard—why should they?— for their wives and children are far away, and safe.'

His face hidden by the night, Odysseus narrowed his eyes and pulled thoughtfully at his dark-red beard. 'The allies,' he asked, 'do they lie amongst the Trojans, or apart from them?'

'They lie apart,' answered Dolon. 'But why do you question me? Do you seek to enter the camp?' In a desperate bid to save his own people, the terrified man said quickly, his words falling over each other with eagerness and fear, 'But why run into more danger than you need? Yonder, on the very edge of the camp, lie the Thracians, newly come to help us, all unguarded, with Rhesus, their king. King Rhesus has the finest horses that I have ever seen, swift, and white as snow, a fair prize for any man. Now take me to your ships and leave me bound to wait for your return, and go and try your luck with the horses of King Rhesus, and see whether I have not told you truly.'

'We shall indeed try our luck with the horses of King Rhesus,' said Diomedes, drawing the sword that he had borrowed. 'But as for you, it is best this way.' And he struck off Dolon's head.

From the severed head they took Dolon's cap of ferret-skin, and the wolf's-hide cloak from his body, and his bow and spear; and Odysseus, taking them into his hands, raised them high to Athene. 'The spoils are to you, great goddess,' he said. 'But now lead us safely to the horses of King Rhesus.'

He hid the cap and the wolf-skin and the weapons in the branches of a tamarisk bush, marking the place with an armful of reeds and boughs, that they should know it again; and then swiftly the two of them made their way towards the spot which Dolon had pointed out to them.

When they were come to the place, they found that it was even as he had said, and the Thracians kept no watch, but were sleeping, each man alongside his weapons, with his two horses hobbled close beside him. And in the midst of his men slept King Rhesus, with his snow-white horses, gleaming in the darkness, tethered to the side of his gilded chariot.

'There lies Rhesus, and there stand his horses,' whispered Odysseus. 'Now we must make speed. Do you kill the men, whilst I loose the horses and bring them forth.'

Diomedes drew his sword and, like a silent shadow, striking down left and right of him, he slew all who lay within his reach; and after him came Odysseus, thrusting aside the dead men to make a path for the horses. Twelve men did Diomedes slay, before any woke from sleep, and the thirteenth was King Rhesus himself, as he stirred, disturbed by a sound.

Swiftly Odysseus freed the horses from the chariot rail, fastening them together with their reins, while Diomedes stood, sword in hand, peering all about him through the darkness, ready to strike down any other man who woke.

Odysseus led the horses safely forth from amongst the dead and sleeping men, hurrying them along with blows from Meriones' bow, since he had no whip. Then he whistled to Diomedes in sign that all was well, and Diomedes rejoined him hastily; and, not without a regret for the gilded chariot which they had to leave behind, they each leapt upon a horse and urged it on towards the shore; while behind them they heard shouts and cries of dismay, and the neighing of horses, as the Thracians awoke and found their king and comrades slain.

At the tamarisk bush they pulled up the sweating horses long enough for Diomedes to dismount and seize their trophies and hand them to Odysseus, and then they were away again to the ships, where the others awaited them anxiously, before the trench.

Much praise did the two of them win then, as their friends crowded round to admire the horses, asking eager questions; and, as they crossed the trench and passed through the gateway in the rampart, all the kings and princes of the Greeks felt encouraged, and somehow less apprehensive of the morrow.

Diomedes tied the Thracian horses amongst his own, beside those of Aeneas; and the spoils of Dolon Odysseus set up in the stern of his ship until daylight came and he could offer them fittingly to Athene, at the time when Diomedes sacrificed the heifer. Then he and Diomedes stripped off their clothing and washed themselves clean in the sea, and, after they had made a drink-offering to Athene, who had cared so well for them that night, and drunk a cup of good wine themselves, they lay down to rest, for the little time that remained before the dawn.

VIII

The Battle

IN the morning, very early, Agamemnon called upon all the Greeks to arm for battle. He himself put on his breastplate of bronze inlaid with bands of gold and shining tin and blue enamelling, with a blue serpent writhing up each side, and took up his fine shield, ornamented with bands of tin and blue glaze about a likeness of the head of the Gorgon, a terrible monster, fit to strike dread into the heart of an enemy.

The daylight had brought courage to him, though he still had little hope of victory; yet he was determined to fight his best that day, that he might be an example to all who had sailed at his command for Troy, even as a high king should.

And so the Greeks gathered together before the rampart

and the trench, and ranged against them were the Trojans, led by Hector and Aeneas, and by Polydamas, a noble Trojan who had been born in the same hour as Hector. And Hector seemed to be first here, then there, and everywhere at once, as he went amongst his men with encouragement and orders.

The two armies met that morning in equal battle, and, for all they had been worsted by the Trojans the day before, the Greeks now held their own, and gave no ground. And so it was until midday, while Zeus, the father of all, in everlasting majesty, looked down upon the city and the ships, upon the flashing weapons, and upon slayers and slain alike.

But when the sun had reached its highest and would climb no more that day, the Greeks began to press the Trojans back towards their city, and many, fearing death, turned their chariots and fled away. The Greeks pursued them across the ford of the Scamander, past the high burial mound of Ilus, crowned by a pillar set up in memory of the ancient king who lay below it, and past the wild fig-tree, to the very gates of Troy.

Yet, by the sacred oak at the Scaean Gate, the Trojans rallied, and stood to meet the Greeks' attack, and the Greeks could go no farther, such being the will of Zeus, even as he had promised Thetis. For in the fighting near the gate, Agamemnon was wounded in the arm, and though at first he did not cease from fighting, after a while his arm grew stiff and he could no longer hold his spear. So, calling to the Greeks to hold firmly the ground that they had gained, he mounted his chariot and returned to the ships.

When the Trojans and their allies saw that the high king of Greece himself was wounded and gone from the battle, they took heart, and Hector went about amongst them calling out to them, 'See, the leader of all our enemies is wounded. The Greeks will lack the courage to stand against us now, and Zeus will again give us this day's victory.'

The Trojans shouted in answer, and with Hector at their head, they hurled themselves upon the Greeks, driving them back as far as the burial mount of Ilus. There Odysseus, fighting valiantly, was aware of Diomedes close by, and called to him, 'Come, my friend, stand beside me. Together, and with Athene's help, we shall hold off all Troy. For Hector must not reach our wall again today.'

Diomedes came to him willingly, and the two of them together wrought great havoc amongst the Trojans, and many did they kill. Hector, seeing that no other could withstand them, came himself towards them, leaping from his chariot, with his chosen warriors close after him. Yet so eager was he, that he easily outstripped all his men.

Diomedes saw him coming and said, dryly, 'See now, Odysseus, here comes death and destruction upon us: Apollo's well-beloved Hector himself. Let him come; we shall not fly before him.' He poised his spear and waited, and when Hector was near enough, he flung it straight at Hector's head. Nor did he miss his aim; but the spear was stayed by the stout bronze of Hector's helmet, which had been given him for his protection by immortal Apollo himself. Yet, though unwounded, Hector was all but stunned by the blow, and he hastened back to safety amongst his followers, and there, all dazed, he sank to the ground amidst them. But after a very little time he rose again, mounted his chariot, and was driven away, while Diomedes reviled the chance which had saved Hector from his spear.

'Truly,' he said bitterly to Odysseus, 'bright Apollo loves Hector well and guards him constantly. But one day, if Zeus is willing, we shall meet again; and if on that day there stands by me a god who loves me as Apollo loves him, then Hector had best take care.' And so Diomedes had to content himself with killing lesser men, which he did with a will, while Odysseus fought beside him, and Sthenelus waited close by with the chariot and horses.

Yet as Diomedes was stripping the armour from a warrior he had slain, to give it to Sthenelus, he was seen by Paris, who had climbed out of reach of the spears to the top of the burial mound of Ilus, from where he had been watching, leaning idly against the pillar. Paris chose out an arrow from his quiver, and sheltering behind the pillar, loosed it at Diomedes. The arrow passed right through his foot, pinning him to the ground. At the look of pain and surprise on his face, as for a moment he wondered what had happened, Paris broke into laughter, and coming from his hiding-place, shouted down to Diomedes, mocking him. 'That was a good shot of mine, son of Tydeus, yet I wish it had been even better. Had I hit you in the body, the Trojans would no longer bleat before your spear like goats before a lion.'

Diomedes looked up at him, undismayed but furious. 'You cowardly archer,' he shouted back, 'you prattler with your curled hair, stealer of the wives of better men, if you came down here and faced me, you would find your bow and arrows little use. You boast now because your arrow has grazed my foot, but I take no heed of such small matters. The blows of a coward and a weakling are like the feeble blows of a woman. Not such are my blows.'

Paris laughed at him, yet he did not wait for Diomedes to take vengeance for his arrow, but ran down the farther slope of the burial mound and made his way lightly to where Hector fought, on the far side of the battle.

With Odysseus standing by to guard him, Diomedes sat down upon the ground, broke off the arrow-head and drew the shaft out from his foot. But in spite of his words to Paris, he was forced to heed his wound, and though it grieved him to leave Odysseus with whom he had done so many bold deeds that day, he limped to his chariot, leaning on his spear, and climbed in beside Sthenelus, who immediately turned the horses and drove his friend back towards the ships, greatly distressed for him.

Odysseus, left alone to stand against the Trojans, was torn in his mind as to what it were best to do: to stay and perhaps be slain or captured, or to leap into his chariot and fly to safety. Yet he soon made up his mind, and setting his back against his chariot, prepared to defend himself, for he scorned to fly. And so mightily did he repel the attacking Trojans, who had thought that he would fall an easy prey to them now that he was alone, that before long the Trojan dead lay all about him.

But at last he was wounded by a spear-thrust in the side; though immediately he slew the man who had given him the wound, even as he turned to run to safety after his too daring attack, plunging his spear through the man's back, between his shoulders, while making light of his own hurt. Yet when the Trojans saw that he was wounded, they believed that he could not long resist them, and they fell upon him eagerly. Odysseus, seeing the great danger he was in, shouted aloud for help, in the hope that someone of the Greeks might hear him.

Three times he shouted loudly, and each shout Menelaus heard, where he fought a little way off. He called to Ajax, son of Telamon, who fought beside him, 'That was the voice of Odysseus. He is in danger. Come, let us go to him.' And together they fought their way through the press of men who surged about Odysseus, like jackals around a wounded stag.

When tall Ajax came amongst them, striking about him on every side, with his huge shield slung before him, so that none of their weapons could touch him, the Trojans drew back fearfully, and Menelaus, going to Odysseus, helped him to his chariot, and carried him safely from the fight.

But Ajax, all alone, did great things against the Trojans, putting fear into their hearts, so that none dared stand against him.

On the far side of the battle, Hector fought against old Nestor and King Idomeneus, and Machaon, the physician; and many of their warriors did he cut down, to leave them lying dead upon the plain. Yet the Greeks would have held their own, in spite of Hector's slaughter, had not Paris, guarded safely by his own followers, taken aim at Machaon with an arrow from his bow, wounding him in the right shoulder, so that Machaon could no longer hold a weapon to defend himself. Seeing this, his comrades were afraid that he would be slain, and Idomeneus said to Nestor, 'Good old king, I beg you, take Machaon up with you into your chariot and drive with him to safety, lest he is slain in spite of us. For a physician is worth many men.'

Immediately Nestor, who, old man though he might have been, would not have left the battlefield for any other cause, made his way to where Machaon stood, one hand clasped to his wounded shoulder, the blood trickling between his fingers, and took him up beside him in his own chariot and turned the horses' heads towards the shore, urging them on with voice and whip, out of the press of battle to the safety of the ships.

While Hector fought on against Idomeneus and the men of Crete, Cebriones, a son of Priam and half-brother to Hector, who was Hector's charioteer that day, saw from the chariot, over the heads of the fighting men, to where Ajax fought so mightily amongst the Trojans. 'Brother,' he said, 'I see a warrior who is slaying too many of our men. From the size of his shield, I think he is the son of Telamon. Would it not be well that we should go and prevent him from such mighty deeds? For the Greeks are taking heart from his achievements and our men are being hard pressed.'

Hector was willing, and the horses bore them swiftly through the battle towards Ajax, Hector striking down right and left of him from the chariot as they went, and scattering the Greeks before him.

And Zeus, from where he watched on Mount Ida, mindful of his promise, turned the tide of battle for the last time that day, filling the Trojans with courage and the Greeks with dismay, so that, little by little, Ajax and the Greeks were driven back towards the shore, yet always striving to retain every step they yielded. Right to the trench and the rampart they were driven; and there they sought to make a last stand to hold back Hector.

Ajax, in the forefront of the battle, fought like fifty men, and beside him stood Eurypylus from rugged Thessaly. Yet brave Eurypylus did not fight long at the side of Ajax. For a third time that day, an arrow loosed by Paris at a leader of the Greeks found its mark, and Eurypylus, wounded in the thigh, had to draw off from the fighting, under cover of his comrades' shields, and make for safety. Shouting all the time to the Greeks to rally to the defence of Ajax, he crossed the track over the trench and went through the gateway in the rampart, limping and leaning on his spear, to make his way alone towards his hut, bidding those of his followers who would have gone with him to return to Ajax, who had far greater need of their help.

IX

The Broken Gate

As the chariot bearing Nestor and Machaon neared the shore, Achilles, watching the battle from afar, standing at the high stern of his ship, saw it and knew it for Nestor's chariot, carrying a wounded man. He called out to Patroclus, who came to stand by him.

'Things go ill with Agamemnon's advocates today.' Achilles gave a wry smile. 'Soon, I do not doubt, we shall have them here again, clasping my knees in supplication and telling us their need is very great. I have this moment seen Nestor's chariot, carrying a wounded man. From here, he looked to me like Machaon the physician, but the horses were too swift and the chariot too far away for me to be sure of it. If it is Machaon who is wounded, that will be a great loss to the Greeks, for, from all that I have seen of this day's fighting, by evening there will be much work for both physicians. Go quickly, Patroclus, and find out if it was indeed Machaon, and how badly he is hurt.'

Patroclus immediately leapt down from the ship and set off to do as Achilles asked, running along the shore past ships and huts until he reached the place where the ships from Pylos were drawn up, in that line of ships closest to the sea. He went to the hut of Nestor, and there he found Machaon lying, and old Nestor sitting beside him, refreshing himself with wine sprinkled with cheese and barley meal, which he was drinking from a cup which he had brought from home: a fine cup it was, embossed with golden studs, and four-handled, and on each pair of handles were golden doves.

As soon as they saw Patroclus, Nestor rose to welcome him, bidding him sit and drink with them. But Patroclus thanked him and said, 'I cannot delay, for Achilles has sent me to find out if it were indeed Machaon whom you brought wounded from the battle. I see that it is, and with all my heart I grieve for him and for the Greeks.' When Nestor would have pressed him to stay for a little, he answered, 'I must not, good king, for Achilles will be impatient for an answer to his question, and it would be wise not to keep him waiting.' He smiled with an affectionate and understanding tolerance of the ways of his friend, and added gently, 'You know how easily he can be angered, and for how little cause.'

He would have gone then, but Nestor said, 'Why should Achilles trouble himself which of the Greeks lies wounded? Diomedes is wounded, and Odysseus, and King Agamemnon himself. Machaon here was struck by an arrow from the bow of false Paris. But what cares Achilles for our sorrows? He is concerned only with his own glory. He waits, no doubt, until he sees our ships go up in flames and his anger is appeased, not only by Agamemnon's death, but by the death of all the Greeks, his friends and one-time comrades in battle. When we are defeated utterly, then, perhaps, his cold heart will melt and he will pity us—too late. Oh, Patroclus, that time when Odysseus and I were in Phthia, at the house of King Peleus, before you and Achilles sailed with us for Troy, I remember how eager you both were to go with us. And when he parted from him, Peleus bade his son be always brave and let no man surpass him in fighting skill. But your father Menoetius said to you—I myself heard him say it—"Achilles is the son of a greater man than I, and he will one day rule over lands far wider than any I have ever ruled, and, young as he is, he is a finer warrior than you will ever be; but you, my son are the elder and the wiser. Never forget that, and be ready always with guidance and good counsel, for he loves you well and will pay heed to you where he would listen to no man else." Have you forgotten your father's words? Even now, when all is almost lost for us, speak to Achilles and plead with him for all our sakes, and we may yet be saved. If he will not fight himself then let him send his men to help us, for they have held many days from the fighting and they will be fresh, while we have fought long and are weary, and defeat and death seem very close to us.'

Patroclus, much moved, said, 'Good king, I will do all I can for you, you have my word on it.' And he hurried away westwards along the shore, back towards Achilles' ships.

But when he drew level with the ships of Odysseus,

before which stretched the assembly place, there he met with Eurypylus, alone, limping from the battle with Paris's arrow-head in his thigh. When Patroclus saw him, streaked with the sweat and dust of battle, the blood streaming down his leg to the ground, he cried out with pity, 'You, too, Eurypylus! What an ill fate the gods have sent us. How many good Greek warriors will lie still and cold tonight, robbed of their lives by Trojan weapons.'

'Good Patroclus,' gasped Eurypylus, 'help me to my hut, for I can walk alone no farther.'

And though he knew Achilles would be waiting for him with steadily growing impatience, Patroclus did not hesitate, but turned aside, and with an arm about him, supported Eurypylus to his hut. There, a servant saw his master coming, wounded, and hastily spread rugs upon the floor for him, and fetched a basin of warm water.

'Cut the arrow-head from my thigh for me, Patroclus,' said Eurypylus, 'for you have skill in such matters, and our two physicians can give me no aid, for I saw Podaleirius in the thickest of the fighting, while good Machaon is himself wounded and needing a physician.'

So Patroclus cut the arrow-head from his flesh and washed the wound clean with gentle hands, and bound it with strips of linen, talking to Eurypylus lightly all the while, that he might think of other things; and afterwards stayed beside him until he was eased a little from his pain.

But while Patroclus had been speaking with old Nestor, the Greeks had been forced to retreat within their wall, shut fast their gates, and prepare to defend their ships and huts from the elated and triumphant Trojans.

But though the Trojans might exult in their triumph, they could in no way get their chariots across the trench and over the rampart, and when the first flush of their victory had cooled, they found that they could do no more than fling stones and shoot arrows over trench and rampart, and shout taunts at the defenders. Hector, frustrated, drove everywhere amongst his men, urging them to cross the trench, exhorting and upbraiding, wild at being thwarted. Yet not even his own horses, for all his lashing, could cross the wide trench, but held back on the brink, neighing shrilly, while Hector chafed at this setback to his final conquest.

But Polydamas came to him, and Hector paused to listen to his words; for though Polydamas was a lesser warrior than Hector or Aeneas, his counsel was ever much respected. 'It is folly,' said Polydamas, 'to seek to drive our horses across the trench. At the top it is too wide for them, and at the bottom it is narrow, and set with stakes, besides. And even should Zeus, by his favour, permit our chariots to cross the trench, how could the horses drag them over the rampart? No, good Hector, let us leave our chariots with a single man to guard each one and to hold the horses, while the rest of us, on foot, leap across the trench and climb the rampart. Where Hector leads, all the Trojans and their allies will follow.'

Hector was pleased by the advice of Polydamas, and immediately ordered all his lords and the leaders of the allies to alight from their chariots and lead their men on foot. Then he divided them all into five companies, setting each company a portion of the wall to attack.

One company he and Polydamas led against the

mid-most part of the wall. This company was the most in numbers, and in it were the bravest of the warriors and those most eager to enter the camp. With them went Cebriones, for Hector knew his half-brother's courage and skill too well to leave him with no more to do than guard a chariot and horses.

The second company was led by Paris and Agenor, one of the sons of Lord Antenor, Priam's good counsellor; for, though Paris was not loved by the Trojans, and though such as he was more aptly armed with a bow and arrows than with spear or sword, he was a great prince and King Priam's son, and it was fitting that he should be called upon to lead.

Helenus and Deïphobus, both younger sons of Priam and Queen Hecuba, commanded the third company; while the fourth was led by Aeneas, and with him were two other sons of Antenor. Over the fifth company Hector put Sarpedon and Glaucus, for the two kings of Lycia were the most important of the allies.

Then, shouting their defiance, with weapons ready and brandished, the Trojans leapt across the trench or scrambled over the stakes and sought to climb the rampart beyond; while from the top of the rampart the Greeks flung down stones upon them. They dropped great lumps of broken rock, which they could hardly lift with both their hands, upon the Trojans' heads; and pelted them with smaller stones from slings.

But the Trojans were many and determined, as they swarmed before the rampart, digging away the piled-up earth, rolling stones and rocks aside to uncover the supporting baulks of timber; and these they tore at, that they might make a breach to enter by. And none were more determined in their attack than the men led by Hector and Polydamas and young Cebriones.

Ajax, son of Telamon, and Ajax of Locris went up and down upon the rampart shouting encouragement to the Greeks, taking a hand here or there, wherever the assault was strongest; the tall son of Telamon lifting up great stones, too heavy for any other man to raise unaided, and dropping them down upon the enemy, and Ajax of Locris leaning from the rampart to thrust with his spear at any Trojan who had found a foothold on the exposed wooden supports.

Yet for all the Trojans' furious assault, the wall still held, and the Greeks, not yet entirely despairing, fought with high courage and unflagging determination, for they were now, and for the first time, defending their ships and their own possessions: the huts which had been their homes for almost nine long years, their slaves and cattle, their supplies and stores of grain, and the hoards of booty which were hidden on the ships.

If Hector's attack was the most to be feared, yet not far behind it in persistence was the attack of Glaucus and Sarpedon. In the very forefront of their company's assault, brave Sarpedon called out to his fellow-king, 'Glaucus, why is it that we two are held in honour above all our people? Why are our cups always filled first, and why is it to us that the choicest portions of the meat are always given? Why are we held in reverence above all men in Lycia? Is it that we may prove ourselves to be no better than any others? Come, let us so bear ourselves that our people may say of us, "Truly, they are fine men, these two

kings of ours who rule in Lycia. They eat of the best in the
land, and drink the richest wine; but so, too, is their might
above all other men's, and they excel in deeds of war." Oh,
Glaucus, my dear friend, had I my will, and all eternity to
live, then would I never again go out to battle, nor would
I let you go forth to cruel war. But seeing that things are as
they are, let us go forward bravely, whether we win glory
for ourselves by our deeds, or whether we win glory for
another by our deaths.' So saying, Sarpedon renewed his
attack, and Glaucus with him, and after them pressed the
army of the Lycians.

The king of the men of Athens, who held that part of
the wall where they attacked, was afraid of their might,
and he called out to Ajax, son of Telamon, who stood by
Teucer on the wall, a short way off, asking for his help.
But such was the din of battle, the shouting of the attackers
and the cries of the wounded, the clash of shields and the
hammering of stones upon the gates as the Trojans tried
to break them down, that Ajax did not hear him; so he
sent one of his men, though he could ill spare him, running
along the walls to Ajax and his brother.

Immediately he heard of the Athenians' plight, Ajax,
with a shout to Ajax of Locris to fight for all three of
them while he and his brother were gone, set off with
Teucer, to find the Lycians clambering up the ramparts,
and the men of Athens in confusion.

Teucer fitted an arrow to his bow, and Ajax, his huge
shield not now held to shelter his brother, but slung at his
back, took up a large jagged rock and flung it at a Lycian
warrior, one of the chosen comrades of Sarpedon, shatter-
ing his helmet and killing him instantly. Teucer's first

arrow struck Glaucus in the arm, so that he had to drop back from the rampart to the bottom of the trench. Blood pouring from his arm, he climbed the farther side of the trench and made his way towards his chariot, that his charioteer might bind up the wound for him.

But Sarpedon, half-way up the rampart, found a firm foothold and thrust upwards with his spear, killing the man who held that stretch of the wall directly above him. Then, before any other man could come to take the dead man's place, Sarpedon, tugging mightily with all his strength at a large baulk of timber, dragged it down, making a narrow gap in the rampart.

Yet before he could take advantage of the opening and leap within the wall, Ajax and Teucer came against him. Teucer loosed an arrow which struck him on the breast-plate but did not pierce it, and Ajax thrust a spear against his shield and forced him downwards. But he was un-daunted. He turned and called out to his men, 'Come, my Lycians, do not leave me to leap the wall alone and fight a path unaided to the ships. Come with me, all of you. The more men who set their hands to it, the easier will be our task.' And the Lycians crowded about him, clambering up the rampart to the breach that he had made.

But to brave Sarpedon was not to be the glory of being the first within the wall, for Ajax and Teucer between them rallied the defenders, and they stood, a strong line, all along the top of the rampart at that place, raining down stones and arrows upon the Lycians, and forcing back with their spears any they could reach.

Soon it was much the same all along that central portion of the wall which was being attacked, with the Trojans and their allies climbing up the rampart, yet never quite gaining the top, and battering in vain on the strong gates set here and there; and the Greeks valiantly defending the wall that they had built, yet never able to do more than barely to repel the attackers, quite powerless to put them to flight.

And thus it went until Zeus, who had not forgotten his promise made to Thetis, gave great glory to Hector, who stood before the largest of the gates, that which was set in the very middle of the wall. A broad gateway of double gates, wide enough for the chariots of the Greeks to pass in and out from camp to plain, it was built of tree-trunks and bolted with two huge wooden bars. Against this gate Hector and Polydamas and their chosen warriors had hurled themselves time after time, until the timbers had bent and groaned, yet still the gate had held. Then Hector, infuriated and beside himself, snatched up a huge rock torn from the rampart and struck again and again upon the gate so mightily that the bars could not hold against such importunate knocking, and at last gave way. With a shout of triumph, Hector took up his weapons, and thrusting the gates apart, he leapt within, a spear in either hand, his armour flashing in the bright afternoon sunlight and the horsehair crest of his gleaming helmet streaming out behind him. In that moment of his victory, none but the immortal gods themselves could have held him back, and no Greek there dared oppose him. For an instant he stood there alone, terrible to see, then he shouted over his shoulder to his men as he sprang forward, and they surged after him, driving before them the defenders of the gateway.

All along the wall the Greeks leapt down from the rampart and fled towards their ships, that they might protect them with spear and sword and shield and body, now that the wall had failed; while through the gateway poured the Trojans and over the rampart they clambered, and everywhere between wall and ships was rout and clamour and confusion.

The Trojans, thronging after Hector, and already shouting out their triumph cries, did not doubt that they would take the ships and the camp, and drive the Greeks back to the very edge of the sea. But, since their wall of wood and earth and stone was taken, the Greeks drew up, a living wall, before the first line of ships. Man beside man they stood, shield touching shield; the bravest in the front rank, their spears bristling in their hands.

In the centre, opposite the broken gateway, were drawn up the forty ships of the dead king, Protesilaus, and behind them, in the second row, lay the forty ships of the Locrians; and it was here, before the ships of Protesilaus, where the main attack was made, that Ajax and Teucer and Ajax of Locris took their stand to await Hector's assault. Ajax, son of Telamon, and his namesake stood close together, fighting side by side as they had fought so often before, and behind them and on either side of them stood the warriors of King Telamon, helmets flashing and spears ready. But the men who had come from Locris with the other Ajax did not stand about their leader, for they were men who wore no crested helmets and carried no spears or shields, but trusted in their bows and slings, rather than in close fighting; and now they stood behind the ranks of the men from Salamis, sending their sharp arrows and their stones high over their heads against the oncoming Trojans.

Surprised that the Greeks had rallied, and seeing how, quite unexpectedly, they were ready for him, Hector, meeting with a hail of stones and a rain of flashing arrows, paused in his wild advance towards the ships, and drew back a little, halting his men and calling out to them, 'Be resolute, my friends, and we shall conquer. For all that the Greeks have ranged themselves as a wall against us, with the help of Zeus, we shall break through that wall as we broke through the other.' And he sent word to the charioteers who waited beyond the trench, that they should drive the chariots through the gateway and hold them ready, in case they should be needed.

A little way beyond where Hector stood, from amongst the Trojan ranks, Deïphobus, his younger brother, stepped forward, his shield held before him, meaning to hurl his spear against the Greeks. But Meriones the Cretan marked him, and hurled his own spear more quickly, so that it struck, resounding, against Deïphobus' shield; and for one appalled moment, Deïphobus feared for his life. But the shield held and the spear fell to the ground, the head breaking from the shaft with the violence of its impact.

Meriones stepped back amongst his companions, angry that he had lost his last spear to no purpose; and, since he had no other weapon left, he thrust a path through the ranks and made his way along the narrow lanes between the ships, towards the vessels of the Cretans in the second row, that he might fetch another spear from his hut. As he reached the king's hut, which stood before those of the lesser lords, he met Idomeneus himself, a spear in either hand, returning to the battle after seeing how the Cretan wounded did.

Finding Meriones there, so far from the fighting, Idomeneus was greatly concerned. 'Are you wounded, Meriones?' he asked anxiously. 'Or have you come only to call me to the fighting?'

Meriones smiled at him and shook his head. 'I have but come to fetch a spear, lord. I broke mine, throwing it at the shield of that vainglorious, empty boaster, Deïphobus. I wish that it had killed him, before it broke.'

'There are twenty spears, at least, leaning ready against the wall of my hut. Take one of those, for they are nearer at hand than yours, and I will wait for you.'

Meriones fetched the spear and came running back with it, and together he and Idomeneus hurried towards the fighting.

'At what place shall we join the fighting, lord?' asked Meriones. 'I think that our defence is weakest to the left. Should we go there?'

'When I left the battle, the two sons of Telamon and Ajax of Locris were drawn up in the very centre, facing Hector. They should be a strong enough defence against him. So let us go to where we are most needed, on the left,' replied Idomeneus.

When the Trojans saw the two Cretans approaching to join the battle they attacked them with a will, hoping to slay them both before they had time to strengthen the Greek defence. But they were too slow for Idomeneus, who was as quick and agile as many a younger man, leaping forth from amongst the Greeks to strike down Othryoneus, an important ally, newly come to Troy, one to whom King Priam had promised his daughter Cassandra for a wife, as the price of his aid. Yet he never lived to wed

Cassandra, for he died, slain by the spear of the Cretan king.

Then Antilochus, the son of Nestor, came to join the fighting at that place; and Deïphobus came amongst the Trojans, since he had won no glory against Ajax and his comrades. The battle grew fiercer as the Greeks were heartened by the boldness of Idomeneus; and when he called out a challenge to Deïphobus, Deïphobus did not dare to face him, but went in search of Aeneas, to bid him stand beside him. Aeneas came willingly, and with him came Paris and Helenus, and Agenor, Antenor's son, and all their followers.

When Idomeneus saw them coming, he called out, 'My friends, here comes brave Aeneas against me, a younger man than I. Do not let me have to stand against him all alone.'

Almost before he had finished speaking, Meriones was at his side, and after him came Antilochus, and so they awaited the attack. Fierce was the fighting between them, and in it young Meriones did great deeds, to the pride of Idomeneus. Amongst others, he wounded Deïphobus in the arm, so that Priam's son, in much pain, was led from the battle to where his chariot waited, just within the wall. He mounted into it and was carried back to the city, bewailing his misfortune.

Then Menelaus came from the fighting farther along the line, and Helenus, seeing him and hoping to gain honour by his death, fitted an arrow to his bow. But even as he drew back the bowstring, Menelaus saw him and threw his spear. The flying arrow rebounded from Menelaus' breastplate, but the spear pierced the hand that held the bow, and Helenus darted quickly back amongst his men, hand and bow alike reddened with his blood. Agenor went to him, and taking from one of his followers the sling he carried, with its cord of twisted wool, he bound the wool tightly about Helenus' hand, preventing further bleeding and making a rough bandage to serve until a better could be found in Troy.

Soon after, Hector left the fighting at the centre of the battle, to see how the Trojans fared to the left and right of him. When he came to where Idomeneus and Menelaus were, he found many of the Trojan lords and allies gone from where he had believed them still to be. And on the very left hand of the battle there remained only Paris, encouraging his followers and those of his two brothers.

Anxious and ill pleased, Hector said, 'Most wretched brother of mine, where are all our good comrades who fight in this war for your sake? Where are Adamas and Asius and Othryoneus, who is to wed Cassandra? And where are our brothers, Helenus and Deïphobus?'

Paris's handsome face was haggard beneath his rich-wrought helmet. 'Why do you rail at me now?' he asked wearily. 'I have been fighting ever since you breached the wall. Many other times have I kept from battle, but not today. Adamas and Asius are dead, and Othryoneus is slain by Idomeneus the Cretan. Our brothers are wounded and returned to the city, but, by the grace of Zeus, they are not dead.' Then, seeing Hector's frown and his distress, as he learnt that other, better men had died, while the king of Sparta's false guest still lived and was not even wounded, he made an effort to appease him. With his ready, charming smile he stepped closer, saying, 'But now that you have come to me here, dear brother, tell me what you would have me do. Tell me where the battle is fiercest and where you would have me fight, and I shall do my best for you.'

'The fight is fiercest where good Polydamas and Cebriones stand against the sons of Telamon,' said Hector shortly. But his anger had been turned aside and he went peaceably enough with Paris, back to the fighting at the centre, and did not chide him further.

When Ajax saw Hector returning to the fight, he called out to him, above the noise of clashing weapons, 'Welcome, son of Priam. Come nearer. Do not keep so far away. You thought to take our ships, but you have not won them yet. Soon will come the moment, Hector, when you stand trembling in your chariot, praying to Zeus and to all the immortal gods that your horses may be swifter than hawks, as they bear you from the battle to the safety of your city.'

So Ajax mocked him, and Hector shouted back, 'You speak nonsense, son of Telamon, braggart and boaster that you are. I shall kill you yet—if you stay and do not fly from me.'

And, once again, he led the assault against the Greeks; and battle broke out afresh, with renewed might.

X

The Fighting at the Ships

AFTER Patroclus had left them, the din and clamour of the battle reached to where old Nestor rested in his hut, in company with the wounded physician, and he listened to it, much disturbed.

'The noise sounds close,' he said to Machaon, 'closer than it should be. May the immortal gods grant that the Trojans have not won through our wall.'

Bidding Machaon stay where he was and not move, because of his wound, Nestor took a spear and went from the hut. Hastening up one of the narrow lanes that ran between his ships, and crossing the wider way which divided the second row of ships from that which lay nearest to the shore, he hurried into another shadowed lane between high hulls towards the first line of ships, with the uproar of the battle growing ever louder in his ears. Even before he saw the fighting, he knew that what he feared had happened, and the Trojans had indeed crossed over trench and rampart and had reached almost to the ships.

For a moment Nestor hesitated, a sturdy, upright, grey-haired figure, standing beside the sheltering hull of some-one else's ship, anger and pity urging him to fling himself, without either shield or breastplate, helmet or greaves, and armed only with a spear, into the battle in defence of the ships and huts; while prudence and duty counselled him first, at least, to go and arm himself and acquaint Agamemnon, as leader of the army, with the way things were going. For, Nestor reflected, Agamemnon, having left the battle wounded, long before the Trojans reached the wall, might well still be ignorant of the plight of the Greeks. As always with good Nestor, duty and better judgement won, and he hurried back towards the shore and the third line of ships, where Agamemnon had his huts.

But when he came to the broad way between the third and second rows of ships, he met the three wounded kings, Agamemnon, Odysseus, and Diomedes, coming, all to-gether—Diomedes limping, and each of them leaning on a spear—to learn what they could of the battle. For all of them, having, like Nestor, their huts behind the row of ships nearest to the shore, had heard and been made un-easy by the sudden nearness of the sounds of fighting.

Seeing Nestor hurrying towards them, Agamemnon called out, 'Why have you left the battle, worthy Nestor? Surely Hector has not made good his boast to break down our wall? Such a thing could not be possible unless all the

gods were with him. Or are there yet others amongst the
kings and princes of the Greeks who, like arrogant, spiteful
young Achilles, refuse to fight any longer under my
command?'

'It is true, great king, our wall is indeed broken down,'
replied Nestor, 'and Hector and the Trojans have driven
us back even to the first line of ships. We must take counsel
quickly as to how best we can help in the defence.'

Odysseus and Diomedes exclaimed in distress at his
words, but Agamemnon, pale and dismayed, said, 'Truly,
the immortal gods are against us and would destroy us
utterly. The Greeks have now no hope of ever taking Troy.
They must endeavour instead, as many as may, to save
themselves and their ships. Come now, let us, whose ships
lie in the line nearest to the sea, have them dragged down to
the water immediately. There let us anchor them just off
the shore until tonight, and then, under cover of darkness,
let us set sail. At sunset—if then by the grace of Zeus the
Trojans leave off fighting—those whose ships lie farther
from the shore may drag them down to the sea and set sail
with us. For when all things are against one, and even the
immortal gods stand with one's enemies, there is no shame
in flying from utter ruin.' So he spoke, thinking in his heart
that at least his ships and all the treasure they held, and
those of Menelaus, which lay by his, would be safe; and
reflecting further that, if he sent word at once to Menelaus
and bade him call his men from the fighting and make
ready to embark, Menelaus, however unwilling, would
obey him, as he always did; and so, come what might,
the two sons of Atreus would see Greece and their homes
again.

After the first shocked instant of silence that followed their understanding of Agamemnon's proposal, it was Odysseus who spoke before the others, and for once he did not stop to choose the words he used. 'By all the gods, King Agamemnon, would you have us basely abandon our friends? You should have led an army of cowards and little men, not such as we are who have fought in the land of Troy for the sake of the sons of Atreus for all but ten years. You had best take care that no more than we, who are beside you now, hear you speak words which no great king should ever be heard to utter.'

Agamemnon glanced at Nestor and Diomedes, seeking support and finding none, then he looked quickly down to the ground at his feet, unwilling to meet again the angry scorn in Odysseus' eyes, old Nestor's expression of pained and stern disapproval, and the contemptuous stare of young Diomedes. He said, as composedly as he might, 'You speak with passion, Odysseus. If you believe that, even hard pressed as they are, the Greeks will not wish to sail from Troy, why then, I will not command them to do so against their wishes.' He paused, and no one spoke. Recovering his self-possession, he looked up at them. 'You have heard my counsel, the best that I can offer. Now gladly will I listen to the counsel of any other man, if he can offer advice that is found more acceptable than mine.' He waited for their answer, outwardly aloof and a little condescending, in the manner of a great king, and inwardly resentful and afraid.

Diomedes looked at Nestor, but saw only a worried frown on his thoughtful face, then at Odysseus, and saw that he was still angry, so, after a moment or two, he spoke.

'I am amongst the youngest of the kings and princes of the Greeks, but since neither Nestor nor Odysseus—both better men than I—has spoken, I shall offer my counsel. I say that there is nothing left for us to do save arm ourselves and go to where our comrades fight, holding back Hector from the ships. Those of us who are already wounded, would do best not to join the fighting, lest we bring more affliction upon our men, by our deaths or capture. But we can at least give them encouragement and exhort them to fight bravely; and the presence of the leader of us all will be some comfort to them. And if—which may the immortal gods forbid—the Greeks should be doomed to perish here, between the Trojans and the sea, why then, we, too, can perish with our friends.'

Nestor and Odysseus received his words with approval, praising his courage; and Agamemnon, seeing there was no help for it, added his assent. And so the four kings hastened to arm themselves and join the Greeks, who fought before the first line of ships, still valiantly holding off the Trojans. There they went, amongst the warriors, urging them on to mightier deeds, and encouraging them greatly.

Before the ships of Protesilaus, where the fighting had all the time been fiercest, and where Hector led the renewed assault, with Paris close beside him, and he and Ajax taunted one another, Hector followed his words with weapons, stepping forward to fling his spear at Ajax, striking him, but failing to pierce his armour.

Before Hector could draw back amongst his men, Ajax bent and took up one of the huge stones upon which the keels of the ships were resting, and which had been dislodged by the press of men about the ships. This stone he flung at Hector, and it struck him on the chest, passing over the rim of his shield. He staggered, gasping for breath, dropped his spear, twisted about, and fell.

The Greeks, with shouts of triumph, immediately ran forward, but Hector's comrades reached him first. Polydamas, Aeneas, and the two Lycians, Sarpedon and Glaucus, covering him with their shields, fought off the Greeks until he had been borne to safety behind the fighting line. Still senseless, he was laid in his chariot, the charioteer whipped up the horses and drove out through the broken gateway in the wall, and across the plain as far as the bank of the River Scamander, followed by a number of Hector's men, grievously distressed, and fearing that he was dead.

They carried him from the chariot and laid him on the ground, in the shadow of a tree, and taking off his helmet and armour, they poured water over him. Hector revived a little and sat up; then he fell back again, overcome by the effort.

But it was not the will of Zeus that the Greeks should be victorious without Achilles, so when he saw Hector hurt and lying amidst the green rushes on the river's bank, far from where the battle raged, he sent Apollo to him, saying, 'Go now, and put great might into Hector, son of Priam, and be beside him in the battle, that he may have the victory against the Greeks this day.'

Apollo, glad that Zeus no longer forbade him from the battle, went willingly to the help of Hector. He roused him, putting great hope and great determination into his spirit, and great strength into his limbs; and Hector armed

himself and leapt on to his chariot and drove swiftly towards the shore, his elated followers running after him.
And before him went Apollo, the Bright One, wrapped in
cloud, bearing disaster to the Greeks.

At the ships the Greeks had taken heart, hoping that
their most dangerous enemy was slain, and Ajax had done
yet more great deeds. But the Trojans, believing Hector to
be dead, were in despair, and by far the greater number of
them had been driven back from the ships and through the
gateway in the wall, their chariots with them; and they were
fighting just beyond the trench when Hector came amongst
them once again.

The Greeks, seeing him, said, 'It is Hector come again,
whom Ajax slew. The immortal gods are with him or he
would not still be living. Let us go back within the wall, and
once more prepare to defend our ships.' And they retreated
within the wall and made towards the ships, all save the
very boldest, who remained with Ajax and Teucer, Idomeneus and Meriones, to hold back the Trojans at the
trench for as long as they might.

But with the coming of Hector the Trojans were beside
themselves with joy, seeing him alive. And the presence of
Apollo filled them with courage, so that they fought
mightily; yet on the Greeks Apollo cast despair and
terror, so that, at last, even the bravest of them turned and
fled within the wall.

From his chariot, Hector shouted to his men, 'On to the
ships, and leave the spoils till later. Let no man stop to strip
the slain until the day is won. And let no man hold back
from the fighting, or I myself will take his life in that

moment when he falters.' He drove his horses straight at the wall; and after him came the chariots of the Trojan lords and allies, and after them the men who fought on foot; and every man shouted, raising his war-cry.

And immortal Apollo, before them, lightly cast down the rampart into the trench for the length of a spear's throw, making a bridge. And over it the chariots thundered and the Trojans poured, rank on rank, on towards the ships.

The Greeks, once more caught between the Trojans and their ships, and filled, this time, through the all-pervading ill-will of Apollo, by dread and a foreboding of disaster, could only pray to all the gods—in the hope that one at least would answer them—and wait for the attack. Like a wave, the Trojans swept upon them, and the Greeks could not withstand their onrush, but were driven back, close pressed against the tall sides of the ships, so that many of them clambered on the decks and fought from there, leaning down to thrust with their spears at the Trojans in their chariots below.

And yet once again Hector and Ajax met, in the midst of the line of battle, at the ships of Protesilaus, yet neither would give way before the other. Hector called for fire to be brought, that he might burn the ship which Ajax and Teucer defended, but no one could come near enough to fling a torch and yet still be out of reach of the spear of Ajax or the bow of Teucer, who stood high on the deck, never missing his aim. Yet when Teucer set an arrow to his bow for Hector, his bowstring snapped as he was drawing it back, for it was not the will of Zeus that Hector should die that day. Teucer, shaken, said, 'I restrung my bow this morning. This should not have happened. There is indeed a god against us.'

'Then leave your bow and arrows, and take up a spear and come and stand beside me,' replied Ajax. 'If they are to have the victory, let it not be without a struggle.'

But when Hector and the Trojans saw Teucer's broken bow, they cried out, 'See, the gods are with us!' and flung themselves against the ships, so that the Greeks were, little by little, forced back into the narrow lanes between the ships, and along them, to take their stand before the huts at the rear. And there good old Nestor went up and down amongst their ranks, heartening them and bidding them fight bravely.

Ajax alone would not leave the ships, but holding a long grappling-iron in his hands, he leapt from the stern of one ship to the next, thrusting down with the grappling-iron any Trojan who tried to climb upon the decks, shouting all the while to hearten those of his men who still fought between the ships, 'My friends, have courage and give no ground. For there can be no retreat for us, who have only the sea at our backs.'

Yet at last, for all Ajax's untiring efforts, Hector and his followers seized the stern of one of the ships of Protesilaus and held it, while Hector shouted again for fire to be brought. And several, seeking to please him, brought torches and tried to set the ship alight, but Ajax was there to prevent them while he yet could, his spear ready in his hand. Twelve men he killed before the ship caught fire.

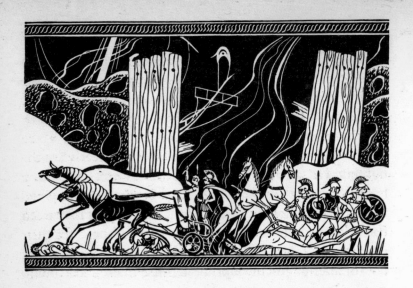

XI

The Parting

As the sounds of battle from around the ships grew
ever louder, Patroclus came out from the hut of
Eurypylus to learn what had befallen. When he
heard that the Trojans were at the ships, he returned, much
distressed, to Eurypylus, and said, 'My friend, I can no
longer stay with you. Things go ill with the Greeks and
I must hasten to Achilles. With the help of the gods, perhaps I can persuade him to forget his anger. If he can be
persuaded by anyone, then surely I can do it.'

He went from the hut and hurried towards Achilles'
ships, hearing the sounds of battle grow fainter behind him
as he made for the most western point of the camp, where
the battle had not reached. But as he came out from behind
the first line of ships, into the open space which stretched
before them as far as the wall, turning to look, he could

see how fiercely the fight was raging farther along, to the centre; and he could see, too, how close to the ships the Trojans had won; and at the sight, tears came to his eyes, and he wept for the sorrows of the Greeks.

Achilles was waiting where Patroclus had left him, at the stern of his own ship, from where he could clearly see the fighting. He saw how hardly pressed the Greeks were, and how many of them had fallen, and in his heart he pitied them and—though he would not have admitted it, even to Patroclus—felt himself in some measure to blame for their present plight. He pitied them, but he was determined that not for anything in the world would he be reconciled with Agamemnon and fight for him again.

He turned as Patroclus approached him, climbing on to the deck of the ship, and marked his tears and read in them a reproach which, he felt, was not unjust, and, defensively, he spoke with a scorn he did not feel, 'Why, Patroclus, what are you weeping for, like some silly little girl? Have you had bad news from Phthia? Is your father dead, or mine? For surely you do not weep for the Greeks, who have brought their own misfortunes on themselves by up- holding Agamemnon when he slighted me?'

But Patroclus knew and loved him too well to fail to perceive the doubts which lay beneath the words, or the carefully hidden pity which his light and mocking tone would have denied. 'I do indeed weep for the Greeks,' he said quietly, 'and it need not make you angry with me. Any man who had not a heart of stone would pity them. Brave Diomedes is wounded, and Odysseus, and even Agamemnon; Eurypylus and Machaon, and I know not how many besides of our friends and former comrades.'

He broke off; but Achilles said nothing, and only looked at him, his lips set firmly and his eyes unyielding. Patroclus exclaimed, 'Oh, Achilles, how can you be so pitiless? I think that good Peleus—who was always kind to me, from the day when I first came to him, young and friendless, an exile in his land—I think that he cannot be your father, nor lovely gracious Thetis your mother; surely, rather, you were born of the cold grey sea and the stark cliffs that tower above it, so hard your heart seems.'

Achilles, moved by his words, came down to him on the deck and would have spoken, but Patroclus, stepping forward, grasped both his hands and cried out, 'Achilles, if you will not go to fight and save our friends, then let me go alone and lead the Myrmidons. Lend me your armour, that the Trojans may see me and think that you have come to the battle, and they will fear for their lives and retire to the city.'

'Not for any reason in the world will I be reconciled with Agamemnon: I have sworn it. And I will not fight again for him. But you know I would deny you nothing, Patroclus.' Achilles smiled. 'Take the Myrmidons and take my armour, and may the gods go with you and give you a great victory.' Impetuously, he went on, 'Drive the Trojans back from the ships and save the Greeks. Show Hector that we are mighty still and do not fear him.' His eyes shone as he talked of fighting, and his voice was eager at the thought of battle. 'Oh, Patroclus,' he cried, 'I wish all the Greeks had sailed for home and all the Trojans were lying dead before us, that just to you and me alone might fall the glory of taking Troy. That would be a day worth having lived to see.' He paused, and his eager smile faded to a wistful self-mockery. He shrugged his shoulders and sighed a little, then said, 'But go now, Patroclus, and come back soon to me: and come back safely. I will call the men together whilst you arm yourself.'

The Myrmidons raised a great shout of joy when they heard that they were to go to battle once again, for they were weary of idling on the shore, and they grieved for the misfortunes of the Greeks.

Automedon, the charioteer, harnessed Achilles' horses to his chariot. Two of them, Xanthus and Balius, were immortal steeds; their sire was the West Wind, they were the swiftest of all horses, and had been given by the gods to King Peleus, on his marriage. Their manes were combed and shining, and their trappings rich, as befitted them: reins and bridles ornamented with ivory, and carved ivory cheek-pieces. With Balius and Xanthus, in the side traces, was Pedasus, who, though no god-born creature like the other two, was yet well matched with them.

Patroclus put on Achilles' armour: the stout leather greaves with their ankle clasps of silver, the engraved and studded breastplate and the helmet with its flowing crest of horsehair: gifts, also, from the gods to Peleus. About his shoulders he slung Achilles' shield and Achilles' brazen sword with its silver-ornamented hilt, and took up two spears of his own, favourites and well-tried; for, besides Peleus himself, only Achilles might use the spear which the gods had given his father.

When the Myrmidons were armed and ready, Achilles stood up before them and cried out to them, 'For many long days you have grumbled that I kept you from the fighting, and boasted of all that you would do if you might meet the Trojans. Now the time has come for you to make good your vaunting. I shall wait for you to return to me with news of a great victory.' And they answered him with a mighty shout, raising high their spears.

Then Achilles embraced Patroclus, and Patroclus mounted into the chariot with Automedon beside him, the long reins in his hands, and, at the head of all the eager Myrmidons, they led the way to where the fighting raged about the ships of Protesilaus.

But Achilles went alone into his hut, and from a coffer which his mother Thetis had given him he took a cup of finest workmanship, from which he would pour offerings to none save only Zeus himself. This cup he now filled with wine, after he had first cleansed his hands with water, and, standing before the doorway of the hut, he poured out the rich crimson stream to Zeus, and prayed to him for Patroclus, his friend. 'Great Zeus, you who dwell on high Olympus and rule all things, hear and answer my prayer. Today I have sent my friend alone to battle. Let Hector learn to his cost that Patroclus is a strong and fearless warrior even when I am not fighting by his side. Grant him victory and give him glory, and when the battle is over, send him safely back to me.' Then, having made his offering and his prayer, Achilles laid away the cup in the coffer and went once more on to his ship, that he might see all that was to be seen from there of how the battle went.

With an eager anticipation, like that which he had felt before his own first battle, he watched and waited confidently for Patroclus to prove himself, before all the Greeks, to be a great warrior in his own right, and worthy of their respect. For ever in the fighting, until that day,

they had fought side by side; and Patroclus, for all his skill
and courage, had always been outshone by Achilles, as any
man would have been by one who was the finest warrior
amongst the Greeks.

Close by the ship which had been fired, where the flames
now rose high whilst the Trojans shouted out their triumph,
Automedon halted Achilles' chariot and Patroclus turned
and spoke to the Myrmidons. 'My friends,' he called to
them, 'let us fight today as we have never fought before,
and win great honour for our lord, Achilles. Let us show
King Agamemnon what manner of man he is whom he
has slighted.' They shouted their war cries, stirred up by
his words, while Automedon lashed on the horses; and so
they fell upon the Trojans who were gathered around the
blazing ship.

Patroclus was the first to strike down an enemy. He
leant from the chariot and flung his spear at the king of the
Paeonians, an ally of the Trojans, who, with his men,
stood close about the stern of the burning ship. The aim
was true, and the Paeonian king fell to the ground, and
never saw his far-off land again.

Utterly dismayed by their leader's death, and believing
at first, from the sight of his armour, that the dreaded
Achilles had come into battle once again, the Paeonians
fled before the Myrmidons' attack, whilst the weary
Greeks raised a cheer at the sight of the help which had
come so unexpectedly to them, and took fresh heart and
set upon their enemies and did great slaughter.

Some of the Myrmidons swarmed upon the blazing ship,
and with sand and sea-water they quenched the flames to

smoke and smouldering spars, then, leaping from the blackened hull, they joined their comrades in forcing back the Trojans to the rampart and the trench.

Menelaus, Antilochus, Ajax of Locris, and Idomeneus flung themselves into the fight with renewed strength, and many Trojan leaders they killed, throwing into confusion their followers, who fled, in wild disorder, towards the wall, clambering up the rampart and flinging themselves down into the trench.

But many Trojan warriors, though they escaped from the Greeks, did not live to reach the rampart, for they were ridden down by their own chariots, crowding towards the gap that Apollo had made, overturning each other as wheel interlocked with wheel and horses plunged, until the gap was all but blocked by the turmoil of chariots and horses. Soon the trench was filled with shattered chariots and broken wheels and men's bodies; while terrified horses, dragging after them splintered chariot poles, galloped wildly over the plain.

Only Hector, whom Ajax, son of Telamon, still sought to kill, bravely stood his ground with a handful of bold comrades and sought to cover the Trojans' retreat. But at last even he saw the hopelessness of staying longer to fight against Achilles' men, who were fresh and unwearied; and he leapt into his chariot and fled, along with his men, through the gap, now clear, save for the broken chariots and the dead.

And after them came the Greeks, with Patroclus at their head, ever shouting encouragement to those behind him; and Achilles' immortal horses, with brave Pedasus beside them, leapt straight over rampart and trench, and on, amidst the fleeing Trojans, to cut off their retreat. And though Hector in his chariot went safely past, on towards the walls of Troy, many of the Trojans were turned back by the overtaking Greeks, and, hemmed in between the trench and the enemy who had caught up with them, they were forced to stand and fight. And so the battle went on, all across the plain.

Ajax, son of Telamon, and Ajax of Locris, come from the ships, fought with a will, weary though they were; and young Meriones, too, seemed tireless.

Meriones met with Aeneas, fighting over the body of a Trojan warrior whom he had slain. Aeneas aimed a spear at him, but Meriones jumped swiftly and lightly aside, so that the spear missed its mark. Aeneas, disappointed, shouted out, 'You are nimble on your feet, son of Molus, and you are, no doubt, a good dancer, but had my spear reached you, you would not have danced again in battle, or anywhere else.'

Meriones laughed at him and taunted in return, 'Conceited though you may be, Aeneas, you can surely not hope to kill every Greek you meet. You, too, are only mortal, and if you were to stand still and await my spear, you would quickly be gone to Hades' land.'

But Patroclus, who was near, heard him and rebuked him in a friendly manner, saying with sincerity, 'Good Meriones, you are brave and skilled enough to have no need of words to strengthen your blows. Words are for the assembly place: blows are for the battlefield. Words waste breath and kill no enemies, therefore it is always best to fight and say nothing.'

Wherever the fighting was thickest, there Patroclus was.

He fought that day as he had never fought before, and put terror into every Trojan heart, killing leader after leader of the Trojans and their allies. Thus did immortal Zeus, father of gods and men, grant one half of Achilles' prayer.

When brave Sarpedon, king of Lycia, saw how many Trojans fell beneath Patroclus' spear, he cried out to his fleeing men, 'Shame on you, you cowards. Will you desert the Trojans for whom we came to fight from so far away? This man, who wears great Achilles' armour, has done much harm to us today. I will meet with him myself, and learn who he is.' And with that he leapt from his chariot, a spear in each hand, and ran towards Achilles' chariot; and with him ran young Thrasymelus, one of his followers, who would not see his lord go alone into danger.

When he saw Sarpedon coming, Patroclus, too, leapt to the ground and stood awaiting his attack. Each of them cast a spear at the same moment. Patroclus aimed at Thrasymelus, and his spear struck the youth in the middle of his body, and he fell dead even as he was running forward. But Sarpedon's aim flew wide and missed Patroclus. The spear passed to one side of him and struck Pedasus instead, and with a shrill cry the horse sank dying to the ground, while Xanthus and Balius reared and plunged at his fall. The reins became entangled and the chariot pole creaked as though it would have snapped; but quickly Automedon drew his sword and cut free the dead trace-horse, and calmed the others.

Meanwhile, Sarpedon threw his second spear; but once again he misjudged his aim, and the spear passed over Patroclus' shoulder. But Patroclus' second spear pierced Sarpedon below the heart, and he fell to the ground almost beneath the hooves of his own horses. His hands clutched at the dust as he tried to raise himself, and with his last breath he called to Glaucus, his friend and fellow-king, saying, 'Glaucus, dear comrade, now must you alone lead the Lycians, since they have lost one of their two kings. Be valiant for my sake.' And then he spoke no more, for Patroclus ran up and pulled the spear from his breast; and so Sarpedon died, far from his own land.

Glaucus, for all he had been wounded in the arm by one of Teucer's arrows, while trying to climb the rampart, prayed to Apollo for strength to do as Sarpedon had bidden him, and he went from one to another of the Lycian lords, seeking to put heart into them. And to Hector he appealed, saying, 'Sarpedon was the noblest of your allies, and the first, save you, over the Greek wall. Help me now to avenge his loss.'

But they were utterly dismayed by Sarpedon's death, Lycians and Trojans alike, and did not fight for long before they fled, Hector with the others; and the Myrmidons stripped the shining armour from Sarpedon's body and sent it to the ships as a rich prize.

But Sarpedon had been a brave warrior and a good king, and well loved by the gods; and Zeus sent Apollo to fetch his body from the battlefield. Like a cloud from Mount Ida, Apollo came swiftly down, took up Sarpedon from the blood-soaked dust, and gave him to the divine brothers, Death and Sleep, who bore him far away to Lycia, that his own kinsmen might bury him in his own beloved land.

Patroclus, overjoyed by the triumph of the Greeks, forgetting that Achilles waited for his quick return, called to Automedon and leapt into the chariot, and pressed eagerly after the flying Trojans, even to the walls of Troy, slaying all he overtook. Followed by the Myrmidons, he flung himself against the walls at that place, near the old fig-tree, where they were weakest. Three times he charged, and three times Apollo thrust him back; until Patroclus, knowing a god to be against him, desisted and withdrew.

Through the Scaean Gate poured in the Trojans, their shields cast aside that they might run the faster, and Hector in his chariot came after them. In the gateway he waited, wondering whether he, too, should seek safety within the city walls, or whether he should go out and meet Patroclus. And Apollo put new courage into his weary heart and new strength into his tired limbs, and he bade Cebriones, his half-brother, turn again and drive the horses once more into battle, and seek out Patroclus. Back towards the on-coming Greeks the chariot raced; but against no lesser warrior Hector raised his weapons, for he coveted the glory of being the one who should slay Patroclus, whose deeds that day had been unmatched; and he sought, besides, to bring hurt to Achilles, most hated and dreaded of all the enemies of Troy, by killing his best-loved friend.

Patroclus saw Hector's chariot nearing him, and he leapt to the ground, a spear in his left hand, and snatched up from the earth a large, jagged stone. When the chariot came near enough, he flung the stone at Cebriones, and struck him on the brow, so that he fell dead from the chariot, head first, the reins still clutched in his hands, and the horses were halted. Patroclus ran towards him as Hector sprang down, and they fought together over the body of Cebriones, until they were separated by the press of men, both Greeks and Trojans, who crowded about them, parting Patroclus, also, from Automedon and the chariot.

Then those Trojans who had followed Hector drew back with him, little by little, in a group, until they were out of range of the weapons of the Greeks; while those Greeks who had been closest after Patroclus gathered about him. Three times Patroclus ran forward alone from amongst his companions, upon the retreating Trojans, and each time he slew many men. And so, in his last fight, he gained great glory for himself.

But as he was running forward alone for a fourth time, Apollo smote him with confusion, so that his senses reeled and he staggered blindly, stunned and bewildered, close upon his enemies. He bent his neck and Achilles' shining helmet dropped from his head into the dust and rolled beneath the hooves of the Trojan horses, and his long hair fell about his face. The shield-strap broke and his spear slipped from his grasp, as he stood unseeing and all amazed.

And while he stood there, alone and unprotected, a Trojan warrior ran forward and struck him from behind, driving his spear deep between the shoulder-blades. Yet such had been the fighting of Patroclus that day, that the Trojan dared not wait to strike a second blow, but snatched out his spear and fled back to his comrades.

In spite of the fearful wound, Patroclus did not fall, but tried, with faltering steps, to make his way back towards the Greeks. Hector, seeing how he was hurt and defence-less, ran forward, and finding him still standing, thrust his spear into his body and felled him to the ground. Standing over him, he laughed in triumph. 'You fool, Patroclus!

You thought that you could take Troy today, but you forgot Hector, whose spear is ever ready to protect Troy's walls. You thought to take Troy and end the war, but instead the vultures shall tear your flesh before the walls you never reached. Unlucky Patroclus,' he jeered, 'whom Achilles sent out to kill me, while he remained in safety by his ships. Much good his love is to you, now that you are dying at my feet, slain by my spear.'

Patroclus, lying before him, twisted on the ground, his hands scrabbling at the spear-shaft, looked up at him, and saw him only through a mist of pain. 'You need not boast of having killed me, Hector,' he said, 'for you only finished what others had begun. The immortal gods, who took away my strength and disarmed me, were before you, and the man who struck me from behind. You were no more than the third. Yet you have doomed yourself to a short life, Hector, by this deed, for Achilles will avenge me.' His voice grew fainter. 'Remember this moment when he comes soon—very soon—to seek you out.' He gave a last shudder before he lay still, and so died.

'Perhaps, for all you know, I may slay Achilles, as well as his friend,' said Hector. But Patroclus could not hear him. Hector set his foot upon Patroclus and pulled out his spear, and spurning aside the body, he leapt forward to find Automedon and the chariot, thinking to kill yet another whose life Achilles prized, and to win for himself the immortal horses; but Automedon, seeing that Patroclus was dead and beyond his help, overcome by horror, turned the chariot, and Balius and Xanthus carried him swiftly out of reach of Hector's spear.

XII

The Dead Warrior

FROM where he was, a short way off, hearing the Trojan's shouts of triumph at Hector's victory, and wondering what they meant, Menelaus came closer and beheld how Patroclus was fallen, and instantly he ran forward.

But as he reached the body, the man who had struck Patroclus from behind came from amongst the Trojan warriors, his spear held ready. 'Stand aside, King Menelaus, son of Atreus,' he said, 'and let me take my spoils. For I was the very first who dared to strike Patroclus today, and his body should be mine to take to Troy to win me honour from the people. Stand aside, son of Atreus, unless you would lie beside him, with my good spear through

your heart.' And he raised his weapon as though he would
have thrown it.

Menelaus did not move. His shield well before him and
his spear grasped in his hand, he stood above the body of
Patroclus to save it for Achilles. 'It is an ill thing,' he said,
'for a man to boast when the gods have given him some
small measure of glory. Go back to your comrades and
taunt me not, lest you do not live to boast in Troy.'

The Trojan flung his spear, but Menelaus caught the
blow upon his brazen shield, and the good shield held firm,
so that the spear did not pass through. Then Menelaus
stepped forward swiftly and thrust the point of his own
spear through the Trojan's throat, and the man fell dead.

Then might Menelaus very easily have borne the body
of Patroclus safely from the battle, for the other Trojans
standing near saw with consternation how their comrade
was slain, and no other one of them dared face the king of
Sparta at that moment. But Hector, finding the horses of
Achilles too fleet for him to catch, and looking back and
seeing that he was likely to lose the body of Patroclus to
Menelaus, turned and hastened towards him, calling out to
the Trojans as he came.

When Menelaus saw how he would be ringed about by
enemies, his mind was torn two ways. 'How can I fly and
save myself,' he thought, 'and leave good Patroclus, who
has died for my sake? Yet if I stay and fight alone against
Hector and all the Trojans, then soon I shall lie beside
Patroclus, and there will be two of us to bring joy to our
enemies.' Yet he stood his ground and would not go, but
instead called out to any Greeks who might be near enough
to hear him, to come to his aid; and most of all to Ajax,

son of Telamon, for he believed him to be not far away.
'Together, just the two of us, Ajax and I,' he thought, 'we
could hold back these Trojans and save Patroclus for
Achilles.'

But Ajax did not hear him, and no one came to his aid;
and at last, for all his courage, he was driven back and had
to leave the body. He turned and fled swiftly, to seek out
Ajax; while Hector triumphantly stripped Achilles' armour
from the body of his friend.

Menelaus found Ajax, his huge shield slung about him,
urging on his men, for they had been greatly dismayed by
the Trojans' recovered strength.

'Come quickly, good Ajax,' cried Menelaus, 'and save
the body of brave Patroclus from the Trojans. His armour
they will have by now, but we can at least try to take his
friend's body back to Achilles.'

At once Ajax went with him to where Hector, the
armour safely stowed in his chariot beside the helmet
which Patroclus had dropped, was drawing his sword.
'This carrion,' he said, 'shall feed the dogs of our city, but
his head I shall have as a trophy to set on a stake upon the
walls of Troy.' He lifted Patroclus' head by the long,
bright hair, and raised his sword, but with a shout tall
Ajax leapt towards him, brandishing his spear, and
Hector gave way before him.

Then Ajax, with Menelaus at his side, stood over the
body of Patroclus, and the huge shield which had so often
sheltered Teucer when he loosed his unerring arrows, now
covered dead Patroclus. And so resolutely and firmly the
two of them stood there, side by side, that for a time there
was no one of all the Trojans that dared oppose them.

But Glaucus the Lycian, grieving for the death of Sarpedon, came to Hector and said scornfully, 'Think now how best to save Troy by Trojan arms alone, for of your allies, the Lycians, at least, will fight for you no longer. For seeing that you left one of their kings alone to die for you, what help from you could any lesser man expect? Sarpedon was a good friend to you, and your guest, and now I cannot even find his body, that I may raise a burial mound above it. No doubt the Greeks have borne it to their ships, that they may show it forth and boast how they have killed a king of Lycia. That man who lies there dead, guarded by but two others, he was the dearest friend of the greatest warrior of the Greeks. If we can take his body into Troy, then we can use it to ransom Sarpedon and his armour. For, if all that I have heard is true, Achilles loved his friend so well that he will pay any price to save his body from dishonour. A fitting burial, at least, you owe to King Sarpedon, the best and bravest of your allies; but you have not the courage to stand face to face with Ajax, for he is a better warrior than you.'

Hector answered him angrily, 'Your wits have left you in your grief, Glaucus, else you would not speak such words. I fear Ajax! I will show you how little I fear him, or any other man.' And in his pride, he put off his armour, and put on the armour of Achilles, which the gods had given to Peleus, and exulting, he stood up before all the Trojans and their allies and cried out, 'To that man who drags the body of Patroclus into the city, I will give half the spoils of this day's fighting, and he shall have equal glory with me today.'

The Trojans cheered his words, and with a great shout

they surged forward against Ajax and Menelaus, until the two were likely to be overwhelmed, for all their courage and determination.

'My friend,' said Ajax, 'in a very little while will the vultures and the ravens and the dogs of Troy be tearing not only the body of Patroclus, but our bodies, too, unless help comes to us, and speedily. Call loudly, Menelaus, for your voice carries far, and bid someone come to us.'

Menelaus cried out with a loud voice, and Ajax of Locris heard him and came running, and after him Idomeneus and Meriones and several others. They formed a ring about the body of Patroclus, and stood, shield touching shield, and faced the Trojans on all sides. And so they fought, for a long while, until some were fallen, and all those who still stood were weary beyond measure; while ever around their wall of shields, the number of the Trojan dead grew greater.

Through all the afternoon they fought to save the body of Patroclus for Achilles—Achilles, who, all unknowing, waited on the shore for his friend to return to him; for the fighting was now close about the walls of Troy, too far away for him to see clearly from his ship how things were going.

Meanwhile, as soon as Automedon had found himself no longer pursued by Hector, he had ceased his headlong flight towards the ships and reined in the horses, and would have turned the chariot once more towards the battle. 'Alas,' he cried out, 'that brave Patroclus should be dead and that I should have lived to see him fall and been powerless to help him. But I may at least try to save his body from the Trojans.'

But when Xanthus and Balius learnt that Patroclus was dead, they would not move, either for the lash or for Automedon's coaxing, but stood there, like two steeds of stone, their heads bent low and their long, silken manes trailing in the dust, and large tears dropping from their eyes. And so they stood and wept for Patroclus, who had so often driven them, and would now never drive them again; and they would carry Automedon neither to the ships nor back into the fighting, until Zeus, watching the battle from far-off Olympus, took pity on them.

'Unhappy ones,' he said, 'why did we gods give you, immortal as you are, to Peleus? Was it only that you might share men's sorrows? For indeed, of all creatures that move and breathe upon the earth, there is none more miserable than man. But you shall not fall into the hands of Hector.' And he breathed strength and courage into the immortal horses, so that they raised their heads and shook the dust from their manes, and neighing shrilly, galloped once more into the battle.

Automedon, distractedly, in vengeance for Patroclus, tried to thrust with his spear at any Trojan within his reach, guiding the horses the while with only one hand; but, impeded as he was, for all his zeal, he succeeded in slaying not one single man; though he came near to death a hundred times himself.

At last a comrade, Alcimedon, saw him and called to him to stop. 'Have the gods robbed you of your wits, Automedon,' he asked, 'that you venture with a chariot in the forefront of the battle all alone?'

'Patroclus is dead,' replied Automedon wildly, 'and I would strike a blow in vengeance for him. Save only he and I, there is no man, besides our lord Achilles, who can drive Balius and Xanthus. But you have some skill with horses. Take the whip and the reins, good Alcimedon, so that I may dismount and fight.'

Alcimedon did as he was bidden, and Automedon leapt eagerly from the chariot, with spear and sword, crying, 'Keep the chariot close behind me as I fight, Alcimedon. Let me all the time feel the breath of Balius and Xanthus on my neck. We must not be parted in the press of battle, for there are many who would wish to take from us the fine chariot of our lord Achilles and the immortal horses.'

Indeed, he spoke truly, for, from where they fought about the body of Patroclus, Hector, seeing the chariot approaching, said to Aeneas, 'Look, my friend, yonder is the chariot of Achilles with the horses which the gods gave to old King Peleus. No more than one man guards it, while he who holds the reins seems to me to be unpractised in managing the horses. The immortal horses should fall easily to our hands if we go together against them. Two such weaklings as those men of Achilles' will never stay to face Troy's two greatest warriors.'

Aeneas laughed, as though the horses were already won, and together he and Hector made their way towards the chariot; and with them went two young Trojans, Chromius and Aretus.

Automedon looked up and saw them coming, the two greatest warriors amongst the Trojans, and he feared for his life and the safety of Achilles' horses, and raised his voice in a shout for help to those Greeks who stood around the body of Patroclus. Yet he did not wait for their aid, but instead, with a prayer to Zeus, he took aim with his spear and hurled it with all his might against Aretus, so that it passed right through the young man's shield and on into his body, and he fell dead. Immediately, Hector cast his spear at Automedon, a well-aimed throw which would have killed the charioteer had he not marked the spear as it came and crouched low to the ground before it reached him, so that it passed harmlessly over his head and struck quivering in the ground behind him.

Upon that the three Trojans ran forward with drawn swords and fell upon Automedon, and he would indeed have been slain in that moment, had not Ajax, son of Telamon, and Ajax of Locris come hurrying towards him, having heard his call. So fierce was their attack, that Hector and Aeneas and Chromius drew back, leaving their dead comrade.

Automedon stripped the armour from the body of Aretus and flung it into the chariot. 'One man at least have I killed in vengeance for Patroclus,' he said. 'And though he was far from being his equal, I have eased my heart a little for the loss of a good comrade.'

Seeing that Achilles' horses were not to be lightly won, Hector and Aeneas returned to the fight about the body of Patroclus with renewed strength, for it irked them that they should be held off from their spoils for so long by so few. And though Ajax, son of Telamon, and Ajax of Locris came swiftly back to the fight, the Greeks were sorely pressed by the Trojans' latest attack, and saw that they could not hold their ground for much longer.

Ajax, son of Telamon, then said to Menelaus, 'It is time that someone bore word to Achilles that Patroclus is dead and lying beneath the walls of Troy, for truly, soon will the rest of us be slain, and all for the sake of his friend.'

'I will go,' said Menelaus, 'and find Antilochus, for he runs fast, and bid him go to Achilles at his ships; though I would by far prefer to remain here with you, where I am so sorely needed. But, good Ajax, and all of you, my friends, while I am gone, do not weaken your resolve to save the body of Patroclus from the Trojans. Now, if ever, remember how good a man he was, how kindly spoken and how gentle to all, and do not leave him for the dogs of Troy to tear.' And turning from them, Menelaus fought a way through the Trojans who pressed them round, and the gap left by his going was closed as his comrades drew nearer together.

Over the plain hurried Menelaus, stopping for no man, until he caught sight of Nestor's son, on the left of the battle, urging on his men to fight. Menelaus came up to him, breathless and weary, and gasped out, 'Ill tidings I bring you, Antilochus. Patroclus is dead. Run to the ships and tell Achilles, that he may come and fetch the body of his friend. As for his armour, Hector has taken that.'

At first, for shock and grief, Antilochus could not move or speak; for, out of all the kings and princes of the Greeks, he had ever held Achilles and Patroclus in the greatest admiration and friendship, and now one of them was slain. But when at last he could use his limbs again, with tears he put off his armour and gave it to his charioteer, that he might run the faster; and so, weeping, he set off with speed for the shore.

Menelaus did not delay, but returned immediately to the comrades he had left, and fought his way once more to the side of Ajax, son of Telamon, who said to him, 'I doubt that we can wait long enough for Achilles to come to us. You fought safely through the Trojans to find Antilochus, so do you and Meriones now take up the body and bear it towards the ships, while Ajax, my good namesake, and I cut a path for you and hold off the enemy. We shall not fail you, for we are old comrades and often before have fought side by side.'

So Menelaus and Meriones took up the body of Patroclus and held it high above them, while Ajax, son of Telamon, and Ajax of Locris, with spear and sword, cut a path for them through the encircling Trojans, who, as they saw their spoils being taken from them, shouted with anger and dismay, remembering Hector's promise of all that he would give to the man who brought the body of Patroclus into Troy.

Step by step, the Greeks fought their way back towards the shore, all the time harassed by the Trojans, and menaced by Hector and Aeneas; until, within sight of the trench and rampart, they were almost overwhelmed.

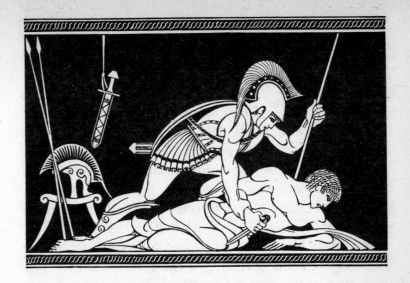

XIII

The Preparation

ANTILOCHUS found Achilles waiting outside his hut, a troubled frown between his brows, for it was fast drawing close to evening, and Patroclus had not yet returned. Moreover, Achilles had learnt from others who still watched that the Greeks were once more being driven back towards the shore; and he feared for his friend. When he saw Antilochus, without his armour, breathless from running, tears wet upon his cheeks, his heart was gripped with dread.

Antilochus, gasping, leant against a pillar of the outer porch. 'I wish it were another man who had to tell you what I now have to say. Patroclus is dead, Achilles, slain by Hector, and Hector wears your armour and boasts how he will have his body, too. Not easily will the Greeks keep it from him.'

Achilles put his hands before his face and flung himself upon the ground, writhing and twisting in his sorrow and tearing at his hair. The captive women, who had been making ready his evening meal, came from his hut and watched him, perturbed and apprehensive. When they knew that Patroclus was dead, they wailed and beat their breasts; for Patroclus had ever been kindly towards them, and much liked.

Antilochus stood there in tears and watched Achilles with pity; until he saw Achilles reach for the dagger at his belt, then he dropped to his knees beside him and grasped his wrists, that, in his agony of mind, Achilles might not take his own life.

And in his wild grief Achilles cried aloud, and his mother Thetis heard him where she sat deep below the sea with her nine and forty sisters, the daughters of blue-haired Nereus, to whom the sailors prayed. Immediately she rose up through the water, and coming swiftly over the sands on her bare white feet, she went to where her son lay and gently touched his head; while up from the sea after her came her sisters, all nine and forty of them, and each one's wailing was the thin sound of the wind upon the waves.

'My son,' asked Thetis, 'why do you weep? For all things are as you wished, and Zeus is giving victory to the Trojans, even as he promised me, when I asked him, for your sake.'

Achilles raised his head from the ground a little way and answered her. 'What good has it done me, mother, that Zeus has granted your prayer? For Patroclus is dead, whom

I honoured above all men and loved more than my life.
He is slain by Hector, who now wears my armour and
boasts of his deeds. Oh, mother, Patroclus is gone, and
what have I left to live for, save to take Hector's life?'

Gently she stroked his hair, saying sadly, 'My most unhappy son, that in your short life you are destined to so
much sorrow.'

'I have lived too long already,' said Achilles, 'since I was
not there to stand beside Patroclus when he needed me.
Oh, that there might never more be strife or quarrelling
amongst gods or men, for it was my accursed quarrel with
Agamemnon that led to this day's grief. But now I shall go
forth to battle once again, and bitterly shall the Trojans
rue my coming. Let me die when I must, as soon as it
pleases the gods, so long as I have lived to see Hector lying
dead.'

Thetis gave a sigh, like the wind that ripples the sea. 'As
you will, my son. Yet do not go forth to battle today, for
Hector wears your armour, that the gods gave to Peleus.
In the morning I shall come again, and bring new armour
for you. Then may you seek out Hector and slay him, if it
is the will of Zeus.'

She rose from beside him and went swiftly to high
Olympus, to the house of Hephaestus, the craftsman of the
gods; whilst, one by one, her sisters slipped below the
waves, their wailing fading on the evening air; and, in its
place, from beyond the rampart and the trench, came the
clash of battle, as the Trojans flung the Greeks back on to
their own defences.

Achilles heard it where he lay, and Hera, queen of all the
gods, sent it into his mind to go upon the rampart and

show himself to the Trojans, that they might see he no longer held from the fight; though because of his mother's command, he would not enter battle that day.

He went upon the rampart and looked down on the fighting, and he shouted aloud with a mighty voice, to encourage the Greeks. For a moment the strife ceased, and the Trojans looked up and saw him there, against the evening sky, with the rays of the setting sun glinting in his long, golden hair, and a great light about his head, as though he had been a god.

Three times he shouted, and a great fear came upon the Trojans, so that they fell back from the wall, and Hector withdrew from the struggle about the body of Patroclus, which once again he had almost won, so that the Greeks were able to take it up. Triumphantly they laid it on a litter and brought it within the camp and set it down with lamentation.

And coming from the rampart, Achilles flung himself across the body of his friend and wept, as darkness fell.

A short way out on the plain, the Trojans gathered in the fading light, to debate what it were best to do; for they were weary from fighting, and afraid, now that it was certain that Achilles would once again take arms against them.

First Polydamas spoke, whose counsel was much respected. 'It were best,' he said, 'that we returned at once to Troy, and tomorrow prepared to defend its walls; since it seems that Achilles, who in the past has made so many Trojan women widows, and taken so many captives, will come once more against us. This time there will be little pity in his heart. He will exact vengeance to the uttermost for his dead friend, and we had best beware of his wrath. Let us return to Troy and guard our city well tonight, and in the morning take our stand upon the walls; and it will be the worse for Achilles or any other Greek who tries to enter Troy.'

His counsel was wise, as his counsel always was, but Hector said angrily, 'You are a coward, Polydamas, and I will not hear your words. No, nor shall any other man of all the Trojans, for I forbid it. We shall camp here tonight, upon the plain, and eat our fill and take our rest; and in the morning we shall again attack the ships. And if Achilles does indeed come once more into the battle, then it is he who had best beware, for I shall be ready to meet him, with my good spear in my hand; and if the gods so will it, he shall end even as his beloved friend, and they can go together to the land of Hades.'

There were many Trojans who praised his bold words and shouted out their agreement with him, scorning Polydamas for a coward. But those others who would have done as Polydamas advised, they kept silent, for they feared Hector's anger.

The Myrmidons bore the body of Patroclus to Achilles' hut, and there they warmed water in a cauldron and fetched out rich unguents and soft woven cloths. And when they had washed away the blood and dust and anointed Patroclus with sweetly scented oils, they wrapped him in a cloth of fine white linen and laid him on his bed in the inner room of the hut, mourning for him all the while they did so.

Achilles, laying his hands upon him, said, 'Here shall you lie unburied, Patroclus, until I have brought you Hector's head. You will not have long to wait, I promise you.'

And all the night Achilles sat beside his friend, unsleeping, his heart within him desolate and aching with grief.

On snow-crowned Mount Olympus, Thetis came to the house of Hephaestus, fashioned of imperishable bronze, all set with stars, which the lame god had built for his own pleasure.

She found him in his smithy, fashioning twenty tripods to stand around the gleaming walls of his hall. Beneath each tripod he had set little wheels, so that they might go of themselves to the feasts and assemblies of the gods, and return again of themselves to his own house, as he wished.

Charis, his fair wife, came forward to greet Thetis and led her to a silver-studded seat and set a footstool at her feet, calling to Hephaestus to come and speak with his guest.

Hephaestus set aside his bellows and laid his tools in a silver chest, and taking up his staff, he limped towards Thetis; and there came with him, to support him, handmaidens fashioned all in gold. So cunningly had he wrought them, that they might indeed have been real maidens.

He took Thetis by the hands and welcomed her, then sat beside her and asked her why she had come to him. 'For if there is any way in which I can serve you, good Thetis, gladly will I do it, whatever you ask of me.'

With sorrow Thetis told him of her son's great grief, and begged that he would make armour for Achilles to take the place of that which he had lost to Hector. 'For my son has but a short life to live—such was his choice—yet I would wish him to have glory in his few brief years.'

'Grieve no more,' said Hephaestus, 'for I shall make for brave Achilles, your son, armour that shall be the wonder of all men who behold it, as I have the skill to do. Yet I would that, when the time comes, I could hide him from death with equal ease.'

With that he limped back to his fire and bade his bellows, twenty in all, to blow their hardest upon the flames; and into the fire he threw strong copper, shining tin, and precious silver and gold; and set his huge anvil on its block and took up his hammer and his tongs.

First, he fashioned a shield in five layers of shining metal, great and strong, with a baldric of silver; and this shield he ornamented with many a cunning design: the sun and the moon and the stars, and figures of beasts, and of men, fighting, dancing, reaping, and gathering rich grapes for wine, a wonder to behold; and all about the rim of the shield was the likeness of the great river, Oceanus.

And when the shield was ready, Hephaestus made a breastplate that was brighter than the blaze of his fire, and two shining greaves and a helmet with a crest of gold.

These arms he gave to Thetis, and she took them up, and like a falcon, she swooped down from high Olympus as the dawn glowed redly in the sky; and as she entered the hut of Achilles, all glorious, bearing the arms, the Myrmidons drew away from her in awe, and she called to her son.

In the inner room of the hut, Achilles, haggard-eyed and pale, looked up from where he lay across the body of his friend, and Thetis set down the arms before him and took him by the hands and raised him. 'My son,' she said, 'for all your sorrow, you cannot bring Patroclus back to life, for he died by the will of Zeus. Let him lie there, and take the armour which Hephaestus has wrought for you alone. No mortal man ever bore finer armour than this.'

His followers turned away their eyes from the splendour of the arms which had been made by a god; but Achilles looked long at them, while his hands clenched at his sides and a fierce light burnt in his eyes. At last he spoke. 'I thank you, my mother, for your care of me, and I thank Hephaestus, also, for his gift. Soon I shall arm myself for battle, and I think that I shall not dishonour this god-wrought gift.'

'Have courage, my son. You will do great deeds today,' she said. And in the next instant she was gone, back to the depths of the sea, and only the armour, lying there before Achilles, showed that she had ever been with him.

Achilles looked once more upon his friend, then went from his hut along the shore towards the place of assembly, calling out to every man he met to join him there.

All the Greeks hurried after him, crying to each other to make haste. Not the warriors alone, and their leaders, who were wont to come, but the pilots, and the helmsmen, who usually remained on their ships, and even the men who had charge of the stores, hurried to the assembly which he called. For Achilles had held himself so long from the war and from the counsels of the Greeks, that all men were curious to hear what he would say, remembering how he had sworn never to take up arms again for Agamemnon in any cause.

Yet now the one thing had happened which, to Achilles,

made all his oaths as nothing: Patroclus was dead and must be avenged.

Odysseus came, and Diomedes, each leaning on his spear, for their wounds were still unhealed, and took their places in the assembly with all the other Greeks. And Achilles sat in his wonted place; for the first time without Patroclus by his side. Last of all, and reluctantly, came Agamemnon, for he had little inclination to face Achilles, now that he had lost his friend.

As soon as all were gathered, Achilles rose, and stepping forward, spoke shortly. 'King Agamemnon, I wish that Briseïs had died on the day we sacked Lyrnessus, before ever I chose her for my share of the spoils. Then would we not have quarrelled over her, and so much sorrow would not have come on all the Greeks. But what has been, has been; and it is over now and done with, and I am ready to fight again. So call the Greeks speedily to battle, that the fewer Trojans shall escape my spear.'

He sat again in his place, and the Greeks made plain their gladness that he was with them and willing to fight once more, by shouting out their approval of his words.

Agamemnon rose to answer him, but in the shouting he could get no hearing, and his words were lost in their cheers. When at last the uproar had died down, he spoke from his place, not coming forward before them. 'My friends, good warriors all,' he said, glowering irritably about him, 'it is fitting to listen when a man stands up to speak, if you do not wish him to confuse his words.' He hesitated, then went on, 'Often enough you have blamed me for my quarrel with Achilles and all that it has cost us, yet these things come from the gods. It was not my fault.

The gods order all things as they please and send what they choose to mortal men. And if I acted foolishly, why, not even the gods are always without folly, as the old tales tell us. Yet, though I am not to blame, to you, Achilles, I shall make amends. I shall send to your ships immediately all those gifts which through Odysseus I promised you two days ago, so that you may see the gifts for yourself and be satisfied.'

Achilles flung out his hand with an impatient gesture. 'Send me your gifts or keep them, as you please. I care not. This is no time to talk of trivial matters. There is a deed yet undone which I must do, so bid the Greeks arm and go to battle.'

Agamemnon scowled, resenting Achilles' scant respect, but quickly Odysseus rose, smiling. 'Come, Achilles, have patience for a little longer. You cannot expect the Greeks to fight fasting. Let us all hearten ourselves and strengthen our courage with good food and drink, before we go out to battle. For he ever fights better who has first taken his fill of meat and wine. And while our meal is being made ready, let our good leader, King Agamemnon, set out the gifts he means to give you here in our assembly place, that we all may see how nobly he has recompensed you, and that you may be gladdened by the sight of all the riches that are to be yours. Let him hand them over to you with goodwill, and then with goodwill, also, let him feast you, Achilles, that the quarrel between you may be done with in the sight of all.'

'That is well spoken, Odysseus,' said Agamemnon. 'I shall do even as you have said.' He turned to Achilles. 'Curb your impatience for a while yet, and accept the gifts

I offer you. Let us make sacrifice to the gods together, and take our meal in friendship.' He quelled his dislike and smiled at Achilles with sullen condescension.

But Achilles exclaimed, 'King Agamemnon, deal with this matter at some other, and more fitting, time. When there is a truce perhaps, or no fighting to be done; or when I am in the mood. Patroclus has been slain by Hector, and you invite me to a feast! If I had my way, the Greeks would fight hungry, and gorge themselves, if they will, tonight, when the dead have been avenged. Yet let them eat if they must, but let them do it quickly. For myself, I cannot think of food or drink, but only of death and vengeance. Patroclus lies in my hut, cold and dead, with his body hacked and torn; I will not eat or drink until I have avenged him.'

Once more Odysseus sought to calm him with lightly spoken words. 'Achilles, you are the mightiest warrior of us all—far better than I in battle—but I am older and have seen more of life, and so must be better in counsel. Pay heed to my advice, therefore. Let us mourn our dead with our hearts, not with our bellies. For, in time of war, men die continually. If things were as you wished them, their comrades who were left would have little chance of eating or drinking, and would speedily fall to the weapons of the enemy, all weak as they would be from fasting.' He smiled and went on, 'No, rather let us mourn our dead and eat and drink our fill as well, that we may be the stronger to avenge our fallen comrades.'

Achilles shrugged his shoulders and wasted no further time in argument, but waited while the promised gifts were brought from Agamemnon's ships and laid before him: seven tripods, twenty shining cauldrons, and much gold. Twelve of Agamemnon's finest horses were then led into the assembly place, and the seven serving women, skilled in handicrafts; and last of all came Briseïs.

Then Talthybius, the herald, brought a boar to the altar in the midst of the assembly, and calling upon the gods to witness his good faith, Agamemnon slew it. Talthybius raised it up, and carrying it down to the sea, he flung it far out into the waves, a sacrifice to the immortal gods.

Immediately he had done, Achilles rose. 'As King Agamemnon says, no doubt it was the gods who robbed him of his wits and sent this trouble upon us.' He spoke in haste, with indifference, adding with more feeling, 'And now, my friends, go to your meal with no more profitless delay, that we may the sooner be at fighting.' And without a glance at the gifts, he turned and walked back to his hut, and the assembly broke up.

The Myrmidons gathered together the gifts and carried them to Achilles' ships, and led away the horses and the women. When Briseïs came to Achilles' hut and saw Patroclus lying dead, she wept and knelt beside him, crying out, 'Dear, good Patroclus, who was always kind to me, will there never be an end to my sorrows? I have seen my husband slain, and my three brothers, and now I lose my kindest friend amongst the Greeks. Unhappy Patroclus, I shall always remember you and your kindness to me, and always mourn your death.'

And when the other seven captive women saw him lying there dead, young, and good to look upon, they, too, pitied him and wept for him; and as she wept, each remembered her own sorrows, and so wept the more.

XIV

The Slaughter

THROUGHOUT the camp the Greeks ate and drank and armed themselves in readiness for the battle that was to come, all save Achilles, and he would touch neither food nor wine. When his followers would have persuaded him, he only said, 'Patroclus is still unavenged.'

Old Phoenix, who had watched them both grow up, stayed with him and tried to comfort him; but Achilles turned away from him, saying, 'Let me be. Ever before, when I prepared for battle, Patroclus himself would make ready a meal for me, and we would eat and drink together before going together to meet the enemy. He lies there dead. Would you have me eat and drink alone?' He put his hands over his face and a sob shuddered through him. 'Oh, Patroclus, no man has ever had a better friend than you.

Never again shall I know another sorrow like this; not though I live to learn from Phthia that my father has gone down to Hades' land, not though I hear that in Scyros my little son is dead. I have known and accepted my doom, that I shall not leave Troy alive; but I believed that you, at least, would go home to Phthia to tell my father how I died, and to fetch little Neoptolemus from Scyros and take him to my own land and care for him for me and teach him battle skill and courage. But it is not to be.' And once more he fell to weeping.

But after a time he raised his head as he heard from all about the camp the sounds of clashing arms and the neighing of horses and the eager shouts of men, as the Greeks gathered under their leaders, confident that they would drive the Trojans back to their city, now that Achilles was come to battle once again. The time for tears was over.

Achilles rose and armed himself in the armour which Hephaestus had made for him: the greaves with their silver ankle clasps and the shining breastplate. He slung the sword and the shield about his shoulders, then, twisting up all his long, yellow hair, he set on his head the helmet with the golden crest. He moved and flexed his limbs, testing the arms and armour, and found them good, and perfectly fitted to him. Then, taking the great spear which had been given to King Peleus by the gods, he went out from his hut to where his chariot waited, with Xanthus and Balius ready harnessed and Automedon beside them; and beneath the flashing, crested helmet, his young face was grim and set, and his eyes were without pity.

Automedon sprang into the chariot and took up whip and reins, and after him Achilles mounted, his armour gleaming like the sun. He stood a moment thus, looking down upon the immortal horses, and then said, bitterly, 'Today, Xanthus and Balius, when the battle is over, bring me safely away. Do not leave me lying dead upon the plain, as you left Patroclus, yesterday.'

Xanthus bent his arched neck beneath the yoke until his long, chestnut mane hung in the dust, and spoke. 'Lord Achilles,' he said, 'this time we shall bring you safely away. But there will come a day when you return not from the battle, for so have the gods ordained. As for Patroclus, he died because the Bright One was against him, and not through our fault or failure, for we are as swift as the West Wind, who is our sire.'

Achilles shook his head and answered quietly, 'Why do you remind me of my doom, Xanthus? There is no need. I know of it already. Nevertheless, I shall not cease from fighting until Patroclus is avenged and the Trojans have had their fill of war.'

Then with a great cry he urged them forward, and after him pressed the Myrmidons, and all the army of the Greeks. And so, for the first time, Achilles went out to fight the Trojans without Patroclus at his side.

From his high throne on Mount Olympus, Zeus looked down on Troy and all the wide Trojan plain; and since this was the day on which his promise to Thetis was to be fulfilled, he no longer forbade any of the gods who would, from joining in the battle. And forthwith Hera, their queen, and Athene with her flashing helmet, brighter than the sun, and lame Hephaestus with them, went from Olympus to watch over the Greeks, bringing to them re-doubled courage; while Apollo of the silver bow and golden Aphrodite, as always, were ready to defend the Trojans, and, once again, dread Ares was with them, dark as a thundercloud.

And so the Greeks, led by Achilles, came from the ships against the Trojans, who were awaiting their attack, drawn up upon the plain; and Hector, wearing Achilles' armour, went up and down along their lines, encouraging his men and their allies to fight boldly. But when he saw Achilles approaching, Hector withdrew himself from the front of the battle; for he knew why Achilles had at last come out to fight again.

Within spear-casting distance of the Trojans, Achilles leapt from his chariot, calling to the Greeks to follow him, and he sprang amongst the Trojans so fiercely that none could withstand him, bringing death to many.

Almost twenty of the finest Trojan warriors he slew in that first short space of battle, showing mercy to none. One of them, who was young and of his own age, cast aside his weapons and knelt and clasped his knees in supplication, and would have pleaded for his life and offered much gold as a ransom. Yet Achilles did not give him even time enough to speak; his sword was in his body and the young man's life was gone, before ever the appeal was made.

Yet it was always Hector whom Achilles sought, looking about for him in the press of battle. But Hector, with prudence, kept afar from him, until that moment when Polydorus was slain.

Polydorus was King Priam's youngest son, a half-brother of Hector. His father had forbidden him to fight, but being young and foolish and eager to prove his courage, he came now to the front of battle, running here and there amongst the fighters, trusting, by the fleetness of his foot, to go unscathed, for he was the fastest runner in Troy.

But Achilles saw him, and swift as he was, Achilles was yet swifter and flung his spear as Polydorus turned to flee from him, so that it struck him full in the back, and he fell to his knees with the spear right through his body.

When Hector saw his young brother dying, he cast aside all caution and strode forward through the Trojan ranks to where Achilles stood. Achilles, looking up, saw him come, flaunting the armour he had taken from Patroclus, and his anger was like a tight hand clasped about his heart. 'Come closer, Hector,' he called out. 'Come closer, that Patroclus may be avenged the sooner.' And he stood waiting for him, his knuckles white about the ashwood shaft of King Peleus' long spear.

'You need not think I fear you, for all you are the greatest warrior amongst the Greeks. I, too, have a sharp spear, as maybe you shall learn.' Hector raised his spear and flung it at Achilles.

But divine Athene breathed upon it gently and turned it aside from its course, and with a shout, Achilles leapt forward and thrust with his spear at Hector. But in that moment Apollo let fall a thick mist all about him, so that Achilles could not see where he should strike. His blow fell wide, and Apollo drew Hector, unharmed, through the mist and away to safety.

Four times Achilles struck into the veil of mist, until at last it cleared, so he could see that Hector was not there, and he cried out in anger, 'This is the doing of one of the immortal gods. No doubt you make rich sacrifice to Apollo before you go amongst the spears, Hector. This time you have escaped me; but there will surely come a moment when no god stands by your side, and in that moment I shall make an end of you.' And forthwith Achilles went to seek out others whom he might slay, until that time when he should again meet with Hector.

He fought ceaselessly and without mercy, until his hands were as red as his spear, and the wheels of his chariot which followed him, and the hooves of the immortal horses, dripped with the blood of those he had slain.

The Trojans gave way before the onslaught of the Greeks, and retreated towards their city; at first in order, making for the ford of the River Scamander, and then, as the Greeks pressed harder on them, in a wild rout.

The ford was not wide enough for a great number of chariots and men to pass over at one time, and in their terror many of the Trojans fled into the deeper water and were drowned, and many drove their chariots over the high banks of the river, so that they sank. Soon the waters of the river were filled with the bodies of men and with the arms that they had cast away as they sought to save themselves.

Achilles set his spear against a tree, and with his sword alone went here and there upon the river's edge, striking down those who would have climbed to safety up the steep bank, clinging to the roots and branches of the willows and the tamarisks that grew there. One after another he cut them down and they fell back into the water, until the river flowed red with their blood.

Many Trojan lords he slew there, beside the river, and

many of their allies. One, amongst so many, was Nastes, leader of the Carians, men who spoke in a strange tongue. Nastes, who had brought his men to Troy, ever went out to battle all decked with jingling golden ornaments, like a young girl. But he lost all his gold when Achilles killed him beside Scamander, and stripped it from his body before flinging him back into the river.

After a time Achilles sheathed his sword and picked out from amongst the others twelve noble Trojan youths, whom he took alive, overpowering them easily, half-drowned and all dazed as they were; and binding their hands behind them with their own girdles, he gave them to his men to be taken to the ships, that they might die on the funeral pyre of Patroclus.

One of those who sought to save himself from the rushing waters of the river was Lycaon, one of King Priam's many sons, the elder brother of Polydorus. Achilles had taken him alive some months before, one night when the youth had stolen out from the city to cut himself the pliant branches which he needed to repair the sides of his chariot. That time Achilles had not killed him, but, for the price of a hundred oxen, he had sold him into slavery in Lemnos. There, at great cost, he had been ransomed by one who had once been King Priam's guest, and wished well to him and to all his family; and so Lycaon had returned to Troy, but eleven days before, to be received in his father's palace with great rejoicing. And now, on the twelfth day after his homecoming, he fell once more into Achilles' hands.

Achilles knew him again as he saw him scramble up the river's bank, unarmed and exhausted, and exclaimed in surprise to see him there. Bewildered, Lycaon stared back at him, standing dripping amongst the rushes and the flowers, dismayed at the ill chance that had given him once again into the power of so mighty an enemy.

'So,' said Achilles, 'a second time we meet, Lycaon, son of Priam. Come, and feel the sharpness of my spear.'

But Lycaon stumbled forward, hands outstretched, for he would have begged his life a second time. Achilles raised his spear, but Lycaon stooped and ran beneath it, so that the spear stroke missed him and the bronze head was buried deep in the ground behind him. Lycaon knelt and grasped Achilles' knees with one hand as a suppliant, while with his other hand he reached back and took hold of the spear-shaft, holding it firmly, so that Achilles might not draw it from the earth too easily.

'Have pity on me, great Achilles,' he pleaded, 'for it must be that I am hated by the gods that they have delivered me into your hands a second time. Spare me now, as you did once before, and take a ransom for my life, for I have eaten bread in your own hut.' He wept. 'Spare me, Achilles, for though we are both sons of Priam, I was not born of the same mother as Hector, who slew kindly, good-hearted Patroclus, your friend. Must I die for Hector's offence?'

But Achilles was moved neither by his tears nor by his entreaties. 'You are too late,' he said. 'Gone is the day when I would have taken a ransom for any son of Priam. They shall all die, all whom I meet with, to pay for the man whom Hector slew.' He gave a bitter little smile. 'Come, my friend, why should I spare you? Patroclus died. He was a better man than you.'

In his cold eyes Lycaon saw no mercy, and he let go of Achilles' knees and unclasped his hand from about the spear-shaft, and with a sob he crouched at Achilles' feet, his hands outstretched amongst the sedges, waiting for the blow which should end his life.

Achilles drew his sword. 'And I, too,' he said, 'I am tall and beautiful, and I am young. My father is a king, and my mother is a goddess. Yet I, too, must die. One day soon there will come a dawn or an evening or a noontide when some man will take my life in battle, with sword or spear or arrow.' He raised the sword and struck Lycaon on the neck, and his blood spattered the flowers and all the green reeds.

Then Achilles threw his body into the river. 'Let the fishes mourn for you, if they will,' he said. 'For your mother will never see you again, to weep over you.'

So it went on, with Achilles slaying all whom he found, as though he could never have enough of killing; until the waters of the river, red with Trojan blood, were choked with Trojan bodies, and still Achilles ranged upon the bank, seeking for more men to slay.

And then at last, in wrath, Scamander, the god of the river, rose up, and in a voice that filled all men with terror, he bade Achilles cease his slaughter. 'Rash and wretched mortal,' he said, 'you have choked my fair stream with dead men and my waters can no longer flow down into the sea. Come, kill no more of the Trojans. They are my people, who, in the past, in time of peace, have offered rich sacrifice to me.'

Achilles, unflinching, answered him, 'I will not cease from slaying the Trojans until they are all fled into their city and until I have met with Hector face to face.'

Scamander rose up in a rushing flood, casting forth on to the land all the Trojan dead, and towering above Achilles, who sought to stay him with his shield, and could not. With the waters swirling about him he grasped the trunk of an elm-tree which grew upon the bank, but with a mighty roar Scamander uprooted the great tree and swept Achilles backwards before him.

Then, because a mortal cannot strive with one of the immortal gods and live, Achilles turned and fled; but the waters of the furious river pursued him across the plain, now washing about his feet and now rising above him as though they would have engulfed him, while he turned and twisted here and there, and tried to reach a place of safety.

But for all his matchless speed, he could not outstrip an immortal, and fearing that he might die there, upon the plain, drowned in the waters of Scamander, with Patroclus still unavenged, he cried out to the other gods for help against the river, and then, since he could run no farther, turned at bay, holding the sword which Hephaestus had forged for him.

Scamander rose higher and higher to engulf him, his torrents roaring mightily. 'Here shall you perish, son of Peleus, you who thought to stand against one of the immortal gods. Neither your arms nor your armour shall protect you, made though they were by a god, but under my waters they shall lie in the slime, and over your body shall I dredge up sand and shingle, so that your comrades will not find your bones when they come to look for them, nor will they need to build any other burial mound for you.'

Then indeed would Achilles have been overwhelmed, but lame Hephaestus, urged on by Hera, came swiftly with flames and kindled fire upon the plain, burning bush and tree, and driving back the waters of Scamander even to his banks. There, on the green margin of the river, the willows and the tamarisks and the cypresses were blackened by the flames, and the flowering rushes and the parsley and the lilies were shrivelled and scorched in the heat, while the stream boiled and steamed, so that Scamander cried out to Hephaestus, 'Cease your burning, for I cannot fight with you, you are too strong for me. Cease your burning, and I will swear an oath that never again shall I help the Trojans, not though I see your flames about their city.'

Hephaestus stayed his fire; and once more Scamander flowed within his banks, down to the sea; and the two gods strove no more together.

But Achilles, delivered by Hephaestus, again attacked the Trojans, slaying ceaselessly, and ever seeking for Hector. While over the plain, towards the city, fled those Trojans who had safely crossed the river, hoping to hide themselves within the walls.

XV

The Vengeance

KING PRIAM came upon the walls of Troy, and from the watch-tower beside the Scaean Gate, he saw his people driven in rout before the spears of the Greeks and he called down to the gatekeepers, bidding them fling wide the gates, that the fleeing Trojans might win through to safety.

Through the Scaean Gate the Trojans poured, with the Greeks coming after them across the plain. But their fear gave them speed, and of those Trojans who had lived to cross the river all passed safely through the gates. All save Hector. He stood before the gateway, watching the on-coming Greeks, and wondering what it were best to do: to save his life and live to fight again, or to stay and face Achilles.

It was Priam, from the watch-tower, who first saw Achilles as he drew near, his god-given armour shining in the afternoon sun, and in fear the old king called down to Hector, 'My son, I beg of you, come within the walls. Do not stay to meet with Achilles. Many of my people has he slain since first he came against Troy, and many sons have I lost to his pitiless spear. Today Lycaon went forth and, against my will, young Polydorus, and I have not seen either return. Oh, my dearest and my best and eldest son, who should be king in my place when I am gone, do not let me have to mourn the death of a third son in a single day.'

Hector looked up at him, and his heart was filled with pity, but he shook his head and set his heavy shield against the wall, and remained where he was, before the gates, alone.

King Priam tore his grey hair and pleaded with his son, and, while he spoke, those who stood near him on the wall moved aside to let Queen Hecuba come by. When she came to the watch-tower, to her husband's side, and looked down and saw Hector standing there alone, she gave a great cry.

'If you are slain, my son,' said Priam, 'who will be left to guard the men and women of Troy? For you are our strong shield and our safety, and if you are killed we shall be left defenceless. Troy will fall to Achilles and the Greeks, and I, your old father, shall be slain, my naked body left for the dogs and the vultures, as though I had not been a king.'

So Priam spoke with tears, yet he could not persuade his son. Nor could Hecuba prevail with him, though she, too,

M 2

wept and said, 'Have pity on me, my child, if on no other. I gave you life: must I now see that life taken from you with my own eyes?'

But Hector, torn between two duties, stood where he was until he, too, could see Achilles coming, in the forefront of the Greeks. Then he almost turned and fled in through the gates, but checked himself, thinking, 'If I go within the walls to safety now, I must face Polydamas, who gave me such good counsel, bidding me order the Trojans back to the city when he knew that Achilles was to fight again: and I would not hear him. He will reproach me, that by my folly so many men have died who did not need to die. How can I go into the city now, to hear some wretch of no account say of me, "Hector, in his pride, has brought ruin on his people"? No, I would rather stay and meet Achilles and either kill him or die gloriously.'

Then he thought how, perhaps for the Greeks as well as for the Trojans, the war had lasted long enough. 'What if I were to lay down my weapons and strip off my armour and go all unarmed to meet Achilles, offer him Helen for Menelaus, and half of the great wealth of Troy, that the war between us might be over and our two peoples be at peace?' For a moment he was eager and hopeful; and then he remembered that it was he who had killed Patroclus, and that there could be no mercy for him from Achilles, though he offered him the whole of Troy and all its riches, for himself alone. 'He would kill me as I stood unarmed before him. For between Achilles and myself there can be no pretty speeches, such as loving youths and maidens toss lightly to one another—fool that I was to think it.' He shrugged his shoulders and took up his shield again.

Moving forward from the Scaean Gate, he saw Achilles running towards him, brandishing his father's long spear, and knew that the time was very close when Patroclus might be avenged.

Then suddenly he thought, 'By all the immortal gods, I do not want to die,' and a great fear came upon him, and he turned and fled along the walls; and with a shout, Achilles was after him, signing to his men to keep away, for Hector was for him alone.

Past the watch-tower Hector fled, and by the old fig-tree, and on to the smoother surface of the wagon track, a short distance from the walls, where the going was easier; and close after him came Achilles. They reached, and passed, the two springs and the washing-troughs of stone, where, in the days of peace, before the Greek ships had come, the wives and women of Troy had brought their clothes to wash. On, on, they ran, as though it were a race that they were running, and for a prize. A prize there was indeed, and it was Hector's life.

Right around the city they ran; and ever Hector sought to run close in beneath the walls, so that the watchers might drop stones and weapons down on Achilles. But Achilles saw his intention, and always contrived to keep between him and the walls, to prevent him. Three times about the city they ran, and each time, as they passed the Scaean Gate, Hector tried to turn aside to escape within; and each time Achilles was there to intercept him and drive him back again on to the track.

But when they came for the fourth time to the springs and the washing-troughs, the immortal gods, beguiling him—for he was doomed—sent Hector's fear from him, and he stood and turned to face Achilles. And in that moment, bright Apollo, till then his constant protector, abandoned him, leaving him, at last, to stand alone.

Breathlessly Hector called out to Achilles, 'I will fly from you no longer, son of Peleus. Here let one of us make an end of the other. But first let us take an oath together that whichever is the victor will respect the other's body. I swear to you, Achilles, that if the gods are with me today and let me take your life, your armour I shall keep for myself, but your body I shall give to the Greeks, that they may burn it as befits the son of a king. Give me your oath to do likewise for me.'

Achilles stood a short way off and leant upon his spear, his breast heaving underneath the shining armour. When he had breath enough to speak, he said, 'You must be mad to talk to me of oaths. There are no oaths made between men and beasts of prey. Wolves do not swear oaths with the sheep within the fold. There can be no oaths between us, Hector.' He straightened up and moved his hand along the spearshaft. 'Now show if you have courage, son of Priam, or whether all your great fame is undeserved, for the time has come for you to pay me for the death of Patroclus.' He raised his spear and flung it; but Hector, watching carefully, crouched down, so that it passed over him and lodged, transfixed and quivering, in the ground behind him.

'Your aim was poor, Achilles. Now may the gods speed my spear and may it find its mark, for of all Troy's enemies, you are the one most to be feared.' Hector cast his spear and his aim was true, for the bronze head struck, ringing, full upon Achilles' shield; but the shield made by immortal Hephaestus held and was not pierced, and the spear fell harmlessly to the ground.

Hector stood dismayed, for he had no second spear; then he drew his sword and stepped aside, ready to fall upon Achilles, who leapt forward to snatch up once again his father's spear: and Athene herself, unseen, put it into his hand. And so the two of them stood close and faced each other, Achilles in the armour which a god had made for him, and Hector in Achilles' armour which the gods had given to Peleus; and each watched the other warily to see where he should strike, for on that next blow would hang the outcome of the combat.

But Achilles knew the weak place in his own armour that he had brought from his father's house and worn so many times: a gap between the breastplate and the cheek-piece of the helmet. And there he thrust his spear, into Hector's neck, and Hector fell to the ground, gasping out his life.

Achilles laughed to see him. 'Fool that you were, Hector. You forgot me when you killed Patroclus. Did you think that I would not remember you?'

Weakly, with fast ebbing breath, Hector whispered, 'I beg of you, Achilles, accept a ransom for my body, that my people may burn it fittingly. Do not leave me for the vultures.'

Angrily and bitterly Achilles answered him, 'Ask me no favours, Hector. Was it not you who would have set the head of Patroclus upon the walls of Troy? By all the gods, I wish that in my hatred I might tear and devour your flesh myself, for the grief that you have brought to me. There does not live the man who could pay me ransom enough for your body. Let your father offer your weight in gold.

He shall never look upon your face again, nor shall your mother lay you on a bier and weep for you.'

The blood dripped from the wound in Hector's neck, and his voice was no more than a murmur. 'Your heart is hard as iron, Achilles, how could I have hoped to move you?' His voice rattled in his throat, his head sank down, and he was dead.

Exulting, Achilles raised the triumph cry and bent to strip his armour from Hector's body; and the Greeks ran forward, rejoicing that the Trojans had lost their greatest warrior. And there were many, seeing Hector lying there defenceless, thrust their sharp spears into his body, who would not have dared to face him while he lived.

Watching, helplessly and with horror, from the walls, the Trojans cried aloud and lamented that their best protection against the enemy was lost. Queen Hecuba shrieked and tore her veil; and it was all that they could do to hold back old Priam, who would have run out through the Scaean Gate in frenzy to reach his son.

Hector's wife, Andromache, was in her husband's house, working at her loom on a strip of purple cloth, adorning it with flowers of every colour. She had just called to her women to set the great tripod and cauldron upon the fire, so that there should be hot water ready for a bath for Hector when he came from the battlefield, for it was nearing evening.

They were hastening to do her bidding at that very moment when she heard the sounds of lamentation from the walls. Her cheeks grew pale and her hand shook, so that the shuttle fell to the floor, and she ran out from the house and through the streets to the wall, and the people made way for her. She looked down upon the plain, gave one cry, and fell senseless into the arms of those about her.

When she came to herself again, she wept and exclaimed, 'To what an ill fate were we born, you and I, Hector. And to what an ill fate has our little son been born, that he is left fatherless while yet a babe.' And so she wept, and her women wept with her.

There were some amongst the triumphant Greeks who were for attacking the city at that very moment, sure that it would fall easily to them while the Trojans were all confounded by the calamity that they had seen and crushed by their great loss.

And though at first Achilles agreed with them, he then said, 'Let Troy be. I have done what I came out to do. Let us return to the ships, for we have had a great victory today. Shall Patroclus lie longer unburied, unmourned and forgotten? I shall not forget him so long as I still live; and even if, in the land of Hades, men forget their dead, yet I shall remember Patroclus, even there. Come, let us go.'

And Achilles, in his hatred, pierced the sinews of Hector's feet from heel to ankle and bound them together with a thong, and made fast the body to the back rail of his chariot, so that the head lay along the ground. He flung in Hector's weapons and his own armour, then leapt himself into the chariot, and snatching the reins and the whip from Automedon, he turned the horses' heads for the shore. He lashed them to a gallop, and Xanthus and Balius went like the West Wind, who was their sire; and so Hector's body was dragged behind the chariot all across the Trojan plain, with his dark hair trailing in the dust.

XVI

The Funeral

BACK at the camp, each leader of the Greeks, with his own men, went to his ships, there to feast the day's great victory. But Achilles had the body of Patroclus, lying on a bier, brought out from his hut and set upon the shore; then three times about the bier the Myrmidons drove their chariots, to pay honour to dead Patroclus. Then they unharnessed the horses from the chariots, and prepared a feast beside Achilles' ships.

Achilles cut the thongs which tied the broken body of Hector to his chariot and dragged it to the bier and flung it at the feet of Patroclus. He laid his hand, still stained with Trojan blood from the day's slaughter, on the cold breast of his friend, and said with tears, 'I have kept my promise,

Patroclus, and brought you Hector's body. Rest well in the land of Hades, for you are avenged.'

Then, though he had no heart for it, he ate and drank with the Myrmidons, for it was more than only the celebration of that day's deeds, it was the funeral feast of Patroclus. But when the others, having eaten and drunk their fill, had gone to rest, he went apart and walked along the shore, where the waves lapped on the sand, and there, alone, he sat crouched in the darkness, his cloak pulled close about him, his head in his hands, and wept.

But he was weary from the long day's fighting, and from his chase around the walls of Troy, and at last, in the silence of the night, that was broken only by the murmuring of the sea, he fell asleep.

Yet his sleep was not untroubled, for while he slept he dreamt. In his dream he saw Patroclus standing beside him, just as he had been in life, with his same smile and his kindly eyes. Looking down at Achilles, he said in that loved and gentle voice, 'You sleep, Achilles, and I am forgotten. While I yet lived, I was ever in your thoughts, but now you think of me no longer. Bury me soon, that my spirit need no more linger on the bank of the dread River Styx, but may pass over into the land of Hades. And give me your hand, Achilles, for one last time, since I shall not come to you again. Never more shall we sit together, apart from all others, and take joy in each other's company. And one final thing I would ask of you, Achilles. You, too, will soon be dead. Do not let our bones be laid apart, but let them lie together in one urn. Many years were we together while we both lived; do not let men part us when we are dead.'

In his sleep Achilles sat up and cried aloud, 'Why do you bid me do for you as I would have done, without your asking? All shall be as you wish. Oh, Patroclus, never before has there fallen on us a sorrow we might not bear together. Come closer to me. Though we have but a moment left to us, put your arms about me and let us find some comfort in the sharing of our grief.' He stretched out his hands, but clasped only the empty air, and woke in tears to find himself alone on the shore, trembling and cold, in the cold light of early dawn.

When day was come, the Greeks brought wood and built a high pyre close by the sea, and the sorrowing Myrmidons carried to it the body of Patroclus, lying on the bier, with all their chariots coming after; and one by one his comrades cut a lock of hair in mourning and laid it on the bier as each took his last farewell.

After all the others came Achilles, who said, 'On the day I left my father's house he promised my hair to the god of our fast-flowing River Spercheius, if I should return safely home. Since I shall never see my land again, my father cannot fulfil his vow, so I shall give my hair to Patroclus, to take with him as a keepsake.' And he cut off his long, yellow hair and laid it in his friend's cold hands.

Then was Patroclus laid high up on the pyre, with painted jars of oil and honey about him; and from his nine hounds, Achilles chose out his two favourites, that they might go with their master down to the land of Hades. And four swift horses, also, he slew, and they were laid about the pyre with the bodies of the hounds, that Patroclus should be well attended on his journey. And lastly, Achilles had brought to him the twelve noble Trojan youths whom he had taken alive from the battle. They came with fear in their wide eyes and reluctant feet; and he cut the throat of each one and cast him on the pyre.

Then, evening being come, with tears he set a blazing torch to the pyre and kindled it, praying to the winds to blow upon the fire. And the flames rose high, blown by the North Wind and the West Wind.

All night the pyre burnt, and all night Achilles went about it, pouring wine upon the ground from a golden bowl and calling upon Patroclus with many sighs. And at that time when the Morning Star begins to herald the approach of dawn, the flames dropped and the fire died down; and Achilles, wearied out, sank to the ground, and, for a short while, forgot his grief in kindly sleep.

When morning was come, he and the other kings and princes of the Greeks quenched the smouldering ashes of the pyre with wine, and gathered up the charred bones of Patroclus. Achilles placed them in a golden urn which he wrapped in a linen cloth and set in the inner room of his hut. 'Let them lie there,' he said, 'until that day when my bones shall be mingled with them. Now let us raise a burial mound for Patroclus. There is no cause to build it high, for it need do no more than mark the place where our ashes shall one day lie together. But when I am dead and gone to join Patroclus in the land of Hades, then let those of you, my friends, who are still left to do the service for us, build it both high and wide.'

They marked off the circle for the burial mound around the ashes of the pyre, and heaped up earth and stones; and when they had done, they would have gone, but Achilles prevented them. 'It is not fitting,' he said, 'that we should hold no funeral games for one who was my friend.' And he ordered a course to be laid out for the races, and had many fine prizes brought from his ships: bronze cauldrons and tripods, and bars of iron, much valued for its scarcity; and horses and slaves, also, for the winners.

All that day Achilles directed the funeral games in remembrance of his friend, commending the skill of the contestants, and, with praise, awarding the rich prizes. But his heart was heavy, and he cared nothing for it all, save that it honoured Patroclus.

In the chariot races the winner was Diomedes, with the horses of Aeneas, and his prize was a tripod and a slave. Ajax, son of Telamon, and Odysseus proved themselves the best at wrestling, yet could neither throw the other; so Achilles gave them each an equal prize. Of the three who could run the fastest, Ajax of Locris, Odysseus, and Antilochus, Odysseus outstripped the others, and won a huge silver mixing-bowl for wine; though, had Achilles been competing, Odysseus would not have won, for no one of all the Greeks could run faster than Achilles.

The arms and armour of King Sarpedon of the Lycians, which Patroclus had taken from him when he killed him, were the prize for that warrior who should be found the best at fighting with a spear. But Ajax, son of Telamon, and Diomedes both fought so mightily against each other that the prize was divided between them. For his skill with the bow, Meriones the Cretan won ten double-edged axes.

To Nestor Achilles gave as a gift a two-handled urn, saying, 'Good old king, you can have no part in the contests with younger men, though once, I do not doubt, you were as skilled and agile as any here. Therefore, since you cannot win a prize, I give you one. Accept this, Nestor, in memory of Patroclus, for you will not see him again.'

And so it went on throughout the day; and in the evening, when the games were done, they parted, and each man went to his own hut, and having supped, slept peacefully.

But Achilles lay sleepless on his bed, tossing from side to side, weeping and longing for Patroclus as he remembered all that they had done together since they had been children: the journeys they had made by land and sea, their coming together to Troy, and all the many times they had fought side by side.

All night sleep kept from him, and, at the first faint light of dawn, he rose and went out from the hut, weary and heavy-eyed with grief. Seeing the body of Hector where it still lay beyond the palisade about the huts, and thinking how it was to Hector that he owed his sorrow, he yoked Xanthus and Balius to his chariot, and once again, in his hatred and misery and anger, he fastened Hector's body to the chariot and dragged it three times about the burial mound of Patroclus; then left it lying there, outstretched upon the ground, and returned once more to his hut in search of rest, his anger a little blunted, but his heart uncomforted.

And every day, at dawn, he did the same, until eleven days were passed.

XVII

The Ransom

ON the twelfth day after Patroclus had been avenged, great Zeus, father of gods and men, put it into the mind of King Priam that he should go alone to Achilles with a ransom for the body of his son. He rose up from the floor where he sat, with hair dishevelled and garments disarrayed, and through his palace—which yet resounded with the lamentation of his daughters and the wives of his sons, who still wept for Hector—calling as he went for Queen Hecuba, he made his way to his vaulted treasure chamber, with its walls and pillars of fragrant cedar wood, where all his royal wealth was stored.

Goblets and vases of gold and copper, gold and silver cups and dishes were there; chests of rich, embroidered garments, woven rugs, bars of gold and silver; and, laid

in wooden coffers, ornaments fit to adorn those who lived
in the house of a king: diadems of hanging golden leaves,
fillets of beaten gold; bracelets, thin, of twisted gold, or
broad and decorated with formal little flowers; necklaces
of golden beads, or silver beads and ivory; golden ear-
rings in the form of coiled serpents, or with hanging
tassels of gold; cunningly wrought brooches to fasten a
rich cloak, and pins of gold and ivory, for long, dark hair.
All these things, and many more, there were in Priam's
treasure chamber.

As he entered, Hecuba, pale and wearied from much
weeping, came to him, asking, 'Why did you call me,
lord?'

'Immortal Zeus has put it in my heart that I should go to
the ships of the Greeks and offer a ransom to Achilles for
the body of our son. Tell me, wife, does this not seem good
to you?'

But Hecuba flung wide her arms and cried out in horror,
'Through our great grief you have lost your wisdom, lord,
that wisdom for which you have ever been honoured.
Gone is it utterly, or you would not think to put yourself
in the power of Achilles, who is cruel and merciless and
has a heart of stone. Remember how many of your sons he
has slain. Remember how he dealt with Hector. He will
have no respect for your age, nor care that you come to him
as a suppliant. No, my husband, let us rather mourn Hector
here in our own house. Do not give Achilles, who has
slain so many of your sons, cause to boast that he has slain
their father, also.'

'My mind is made up,' said Priam, 'and I will go, for I
believe that the immortal gods are with me in this thing.

But if I am mistaken and Achilles kills me—well, I am an old man now, and I shall die gladly if first I have looked once more upon my dear son's face and bidden him farewell.'

He would listen to no more of Hecuba's entreaties, but, opening chests and coffers, he took out the finest of the garments that lay in them, and costly woven stuffs, and golden ornaments. Ten talents of gold he weighed out, and chose two shining tripods and four cauldrons, and a Thracian cup, the most precious that he owned.

Then he called to his sons, Paris, Deïphobus, and Helenus and those others who were still left to him, and bade them take the treasures and lay them in a wagon. And when they would have protested, he grew angry and rebuked them. 'Make haste when you are bidden, worthless ones. I would that you had been slain in Hector's stead. All my good, obedient sons are dead: brave Cebriones, young Polydorus, Lycaon; and Hector, the best of them all. While you still live, idle wretches who are fit for nothing but for dancing and the telling of lying tales. Do as I bid you, and with no more delay.'

They hastened to obey him, though they thought that he was indeed crazed with grief, in that he meant to go alone to the ships of the Greeks. They harnessed mules to a new-made wagon and laid upon it the treasure he had chosen out, and brought a chariot for their father, and his best horses.

Priam called to him Idaeus, his old herald, and bade him drive the wagon on ahead; but when he would have mounted the chariot, Hecuba came out to him, carrying wine in a golden cup. 'Make the drink-offering to im-

mortal Zeus before you go, lord, and pray that he may
send you safely home from amidst our enemies. Ask of
him a sign, I beg of you, that you may be certain of his
favour. Ask that he may send an eagle, flying upon your
right hand, as a sign that the gods are indeed with you in
this thing.'

'That is well thought of, wife,' said Priam. And when he
had washed his hands in clear water, he took the cup and
poured out the wine and prayed to Zeus for a sign; and
then, as he and all the others there anxiously watched the
sky, they saw an eagle with wide, dark wings flying across
the city, on their right; and even Hecuba was cheered a
little by the sight.

Then Priam mounted on to the chariot and took up the
reins and went out through the gateway of his palace, with
old Idaeus going on before him in the wagon with the
ransom.

Down from the Pergamus and on through the wide
streets of Troy they went to the Scaean Gate; and Priam's
sons and the husbands of his daughters followed them,
mourning as for one going to his death. But once through
the gateway and beyond the sacred oak, the young men
turned back, and Priam and the herald went on alone.
Past the burial mound of King Ilus they went and on towards the River Scamander, and close by the ford they
stopped to water the horses and mules.

And there, by the river, Hermes, swift messenger of the
immortal gods, met them in the likeness of a youth.
Idaeus, looking up, saw him come, and said fearfully, 'My
king, there is a young Greek coming towards us. Shall we
flee, taking the treasure with us, or do we stay and implore

his mercy? For though we are two, we are both old, and he is young and armed.'

But Hermes greeted them kindly and asked them whither they were bound; and when they had told him, he said, 'I shall myself conduct you to Achilles' hut, for it is not well that you should be alone and unescorted, bearing such rich treasure with you. Trust yourself to me, and have no fear, good king.'

'Truly,' said Priam thankfully, 'the gods are indeed with me in this, that I should have met with one so kindly and courteous as yourself.'

So Hermes went with them, and when they were come to the ships of Achilles, Hermes caused the wagon and the chariot to pass, unseen of any man, through the gate in the palisade and on into the wide enclosure where the Myrmidons were making ready their evening meal. There he leapt down from the chariot, saying, 'Go quickly in, good king, and ask your boon of Achilles, entreating his mercy in the names of his old father, Peleus, and of silver-footed Thetis, his immortal mother.' And with that he was gone from them, back to snow-crowned Mount Olympus.

Priam came down from his chariot, and leaving Idaeus to hold the horses and the mules, he went alone into the hut of Achilles.

Achilles, having at that moment finished his meal, was sitting, unsmiling, apart from his companions, and only Automedon was near him, to pour more wine for him, or fetch more meat, should he demand it. Straight to Achilles old Priam went, unhesitatingly; and before he could be prevented, knelt in front of him and clasped his knees in supplication, and kissed the hands which had slain his son.

Achilles looked at him with amazement; whilst all his companions in the hut fell silent, watching the stranger who had come amongst them, and wondering who he might be.

'Great Achilles, I beseech your mercy in the name of King Peleus, your father, whose years equal mine, and who, even as I, stands now on the sad threshold of old age. Yet, unlike me, he lives with hope, hope that one day soon he will see his dear son return to him, victorious after many battles. But I, I have no hope to lighten my remaining days. The dearest and the best of all my sons is gone, and I shall never speak with him or hear his voice again. Nor will he ever return victorious to his father's house. Yet one little grain of comfort would it bring to me, in my last years, if I might look upon his face once more and touch him with my hands; and a great solace would it be to me if, in the sad and lonely, unprotected days which yet remain to me, I could remember how his body had been burnt with all honour as befitted the son of a king, and if I could look upon the burial mound which his mourning comrades had raised for him. In the name of your own father and in the name of immortal Thetis, your mother, have mercy, great Achilles, and give back to Priam, that most wretched king, the body of Hector, his son. A great ransom have I brought you for him; do not refuse it.' The tears streamed down his cheeks and he could hardly speak for weeping. He clasped Achilles' hands and said, 'I implore you, have pity on an old man who must humble himself and kneel to one who has slain so many of his sons.'

Achilles, much moved, gently loosed the old king's clasp and put him from him. His eyes were filled with tears and he turned his head away, thinking how his father Peleus would wait in vain for his homecoming. 'I shall never see my father or my own land again,' he thought. Then he thought how he would never again see Patroclus, a far greater grief than any other, and the tears would not be denied. And so they wept together: Achilles for his father and his friend, with the old king crouched at his feet, weeping for his son.

But at last, as Achilles' tears grew less, he was once again aware of Priam, and immediately he stood and raised the old man kindly, saying, 'Your courage is indeed great, King Priam, that you ventured alone amongst your enemies to seek me out. But, come, sit here upon this seat and let us put aside our sorrowing. For of what avail are tears? They cannot bring back the dead. That is a thing which I have learnt.' And he led him to a chair.

But Priam shook his head sternly. 'Should I sit and take my ease, son of Peleus, while Hector lies in the dust amongst his enemies? No. Give him back to me and let me see him. Take the ransom I have brought you, it is a fitting price for the son of a great king.'

For a moment Achilles frowned, then with an effort he forced down his rising anger and answered, 'Do not provoke me with ill-chosen words, good king. I shall give you Hector's body, as you ask, for I think that you came here today by the will of the gods, else could you not have reached my hut unharmed.' He signed to Automedon to follow him, and quickly left the hut.

Outside, he called Idaeus, bidding him go in to his master and rest himself; then he ordered Automedon to unharness the mules and the horses from the wagon and the chariot, and to carry the ransom into his hut. 'But leave a tunic and two cloaks to wrap the body in,' he said, 'lest Priam, seeing his son's wounds, should reproach me, and I should grow angry at his words and do what I would afterwards regret. For I would not dishonour the gods by mistreating a suppliant, and he a grey-haired king.'

He bade the captive women take Hector's body from where it lay, and, in some place apart, where Priam could not see them, to wash it and anoint it with oil and wrap it in the tunic and the cloaks. And when all had been done as he commanded, he himself helped to lay Hector's body on the wagon; and as he did so, in his heart he cried out, 'Do not be angry, Patroclus, if, even in the land of Hades, you hear that I have given Hector's body to the father who loved him. Do not be angry with me, but understand why I have done it, even as you would have understood while you yet lived.'

Then Achilles returned once more into his hut. 'I have done as you asked, King Priam. Your son lies on your wagon. Tomorrow you shall bear him back to Troy. But now you shall eat and drink with me, and for tonight you shall lie safely beneath my roof.'

Priam sat, and Achilles gave him roasted meat upon a platter, and Automedon brought bread to him in a basket; and in silence they ate and drank together peaceably, the old king and the young warrior who had slain his son.

When the meal was finished, it had grown dark, and a slave set torches on the walls, and Priam, rested and refreshed, looked well for the first time at Achilles, all golden in the flaring torchlight, and saw how all that he had heard of his peerless beauty was true, and he thought, 'Truly, rumour has not lied. He is like one of the immortal gods to look upon.'

Achilles, watching, marked how Priam gazed at him, but he said nothing; and it was Priam who broke their silence.

'This is the first time in twelve long days,' he said, 'that I have sat to eat and drink with other men, for I have cared nothing for food or wine since Hector died. As little as I have tasted, so little have I slept for grief. But this evening, son of Peleus, I have eaten and drunk with you, and now I am weary and would sleep.'

At once Achilles ordered beds to be prepared for Priam and Idaeus in the outer porch. 'Forgive it, good king,' he said, 'that I put an honoured guest to sleep outside. But if any leader of the Greeks should come here through the darkness to speak with me tonight—as well might be—he would see you, should you be lying near the hearth, and know you for the king of Troy. And if word reached Agamemnon that you were here, no doubt, to spite me, he would try to make it difficult for you to take with you the body of your son, which I have given you leave to take. But now, before we part tonight, tell me, how many days do you need to celebrate the funeral rites of Hector? For I doubt, with matters as they are, that the Greeks will go out to fight unless I go with them; and I shall keep from battle for as many days as you need.'

Priam, touched by his offer, said, 'I am grateful for your kindness, son of Peleus. May the gods reward you for it. You must know that it is not easy for us, pent up in our city as we are, to venture far from the walls to fetch wood. Give us eleven days, good Achilles. Nine days to mourn for Hector and to gather the wood we shall need for his pyre; on the tenth day we shall burn him and keep his funeral feast, and on the eleventh we shall raise the burial mound above him.' He paused, then added, 'On the twelfth day, if need be, we shall fight again.'

Achilles smiled a little. 'You shall have your twelve days' truce, King Priam, I give you my word on it.' He laid his hand for a moment upon the old king's arm in reassurance, saying to him, 'Now lie down and sleep, you and your herald, and have no fear, for you will be safe beneath my roof.'

And so, amongst his enemies, Priam slept calmly and in peace; but long before dawn Hermes, once again in the likeness of a Greek youth, woke him, saying, 'You came to Achilles as a suppliant, and as a guest he has received you. Him you can trust, for you sleep beneath his roof; yet you can trust no other amongst the Greeks. Do not delay here, close by the ships. A great ransom you have given for your son, but it would be a ransom three times as great that the people of Troy would need to pay to buy your freedom, once Agamemnon knew that you were here.'

Hastily, and in fear, Priam roused Idaeus, and they hurried from Achilles' hut, while Hermes himself harnessed the mules and the horses and, once again, led them safely past the ships and across the plain. At the ford of the Scamander he left them and returned to high Olympus.

As the sun was rising, the wagon and the chariot came slowly along the track towards the Scaean Gate. Cassandra, Hector's sister, watching from the walls, beheld, in the first light of day, her father and Idaeus, and saw that they had returned with him whom they had gone to fetch, and with a loud cry she roused all the people. 'Come, all you men and women of Troy. If ever you welcomed back Hector, living, from the battle, come now and welcome him, dead. For he died to save our city and us all.'

Out from their houses and out from Priam's palace came the people of Troy, crowding about the gates as Priam brought home his son, and with tears and wailing they welcomed him, following the wagon up the wide street to the Pergamus.

In the palace Hector was laid upon a bier, and about him stood the singers to sing his dirge; and one by one the noble women of the household made lament for him.

First spoke Andromache, his wife, with her arms clasped about her husband's head. 'Oh, Hector, ill fated are we whom you have left defenceless, and ill fated is your little son, but most ill fated of all am I, your unlucky wife; for you did not stretch out your arms to me when you were dying, nor speak to me in words which I might have remembered with tears and treasured in my heart for all my life.'

Beside her stood Queen Hecuba. 'You were the eldest and the dearest to me of all my children, Hector,' she said. 'And while you lived divine Apollo, the Bright One, cared well for you. Yet even in your death the immortal gods did not utterly forsake you; for, though I dared not hope to do so, I have seen your dear face once again, and in all honour do you lie in your father's house, and in all honour shall we celebrate your funeral.' But, after that, she could say no more for weeping.

Then Helen came to them, and the women drew aside as she passed and would not look at her, walking slowly until she stood beside the bier, looking down on Hector. 'Kind Hector,' she said quietly, 'of all the brothers of Paris, you alone never reproached me for the trouble that I brought on Troy, and always would you defend me against the harsh words of others. Now that you are gone, in all Troy I have no man to speak for me, save your father, who, with you, alone was kind.' And she drew her veil across her face to hide her tears.

For nine days was Hector mourned; and on the tenth day, when dawn was come, they laid him on a tall pyre and fired it. All day it burnt, and through the night the ashes smouldered; but on the morning of the eleventh day they poured wine upon the ashes and gathered up the dead man's bones and laid them in a golden urn and set upon it stones and earth, raising a high burial mound.

And so, by the compassion of Achilles, was Hector buried, who had slain Patroclus and brought Achilles so much grief.

EPILOGUE

How the War with Troy was Ended

AFTER the death of Hector, the war with Troy went
on, though things were hard for the Trojans without
their finest warrior; and many times the Greeks
carried the battle right to the very walls of Troy.

It was on one such day that the doom foretold for
Achilles fell on him. Leading the Greeks against the hard-
pressed Trojans, he was fighting close to the Scaean Gate,
where Paris, loath as ever to face the dangers of battle,
stood with his bow in the shelter of the gateway. When he
saw Achilles come within bowshot's range, with a prayer
to Apollo the Archer, who had shown himself such a good
friend of Troy, he loosed an arrow. Never had his aim been
surer, and Achilles, whom Paris would not have dared to

face, fell dead. And so was the greatest warrior of them all slain by a man of little courage and no worth.

After a great fight in which Ajax, son of Telamon, slew Glaucus of Lycia, Ajax and Odysseus saved Achilles' body from the Trojans and carried it back to the ships and laid it upon a bier on the shore. There all the kings and princes of the Greeks gathered with Achilles' own Myrmidons to mourn him. Then up from the sea came his mother Thetis with the nine and forty daughters of Nereus, all wailing for Achilles. At the sound, and at the strange sight of the immortal sea-nymphs, a great fear came upon the Greeks, and they would have fled, but old Nestor, in his wisdom, called to them to stay. 'We have no need to fear,' he said. 'It is immortal Thetis and her sisters come to weep for her dead son.' And the Greeks were afraid no longer, but remained to weep with the goddesses; and so they mourned for great Achilles, mortal men and immortal goddesses, for seventeen long days. And on the eighteenth day the Greeks burnt the body of Achilles on a high pyre with jars of sweet oil and honey. The next day they took his bones from the ashes and laid them in a two-handled golden urn which Hephaestus had given to Thetis. And into this urn, also, they put the bones of Patroclus, and above the urn they raised, even as Achilles had asked it of them, a high burial mound on a headland which overlooked the sea.

At the funeral games in his honour, Thetis declared that the god-wrought arms of her son should be given to the bravest of the Greeks. Because they had, together, brought his body safely from the fighting about the Scaean Gate, both Odysseus and Ajax, son of Telamon, claimed them. But Agamemnon, who wished to please Odysseus, awarded the arms to him, and Ajax, disappointed and angry, went out of his mind and fell upon his own sword, and his half-brother Teucer raised a burial mound for him close by the ships of Salamis.

Paris did not survive long to boast of his triumph. Wounded by a Greek archer, he died, later, of his wounds, and Helen, now quite alone and friendless in Troy, and bitterly regretting that she had ever left Menelaus and her home in Sparta, was grateful for the protection of Deïphobus, the brother of Paris, and went to live in his house.

With Achilles and Patroclus both dead, the Myrmidons were without a leader, so Agamemnon sent to Scyros for Neoptolemus, Achilles' young son, without whom, so Calchas the prophet had said, Troy would never be taken. Though he was still a boy, Neoptolemus had inherited his father's courage and much of his battle skill. He came willingly to the war, and willingly the Myrmidons followed him as their leader; and Odysseus gave up to him the arms of his father which Agamemnon had awarded him.

Yet still the war with Troy dragged on, and the Greeks, seeing that they could not take the city by force, decided to take it by a trick. Many say that it was Odysseus who was the author of the cunning ruse by which Troy fell, and indeed, from all else that is told of him, this seems most likely.

The Greeks built a huge horse of wood, all hollow within, and in the body of this wooden horse fifty of their boldest warriors hid; amongst them Menelaus, Diomedes, and Sthenelus, Odysseus himself, and young Neoptolemus.

The Greek army, led by Agamemnon, then dragged the ships down to the water's edge, burnt the huts, and set sail, as if for Greece and home, leaving the wooden horse alone on the beach. But the Greeks sailed no farther than the island of Tenedos, a little way along the coast, and there they cast anchor and waited.

Amazed and joyful that the long siege seemed to be over, and their city free once more, the Trojans poured through the Scaean Gate and across the plain to the shore. There, amongst the ashes of the camp, stood the horse, and the Trojans wondered at it. 'It is an offering to the gods,' said some. 'It is a trap,' said others. 'Let us destroy it,' they cried.

But King Priam said, 'It is a sacred offering. It must not be harmed. Let it be taken into our city.' And he would not listen to the protests of those who would have advised him against such a course.

So the horse was dragged into the city and honoured as an offering to the gods. And there in the assembly place it stood all day, garlanded with flowers, and with petals and green boughs strewn beneath its hooves. And all day the Trojans made merry and rejoiced that the war was over at last, and their city safe.

That night, when darkness had fallen, the Greeks opened the hidden door in the side of the horse and leapt out. Meanwhile, Agamemnon had sailed back with the army from Tenedos, and landing once more on the Trojan shore, hastened across the plain. The Scaean Gates were opened to him by those Greeks already in the city.

The Trojans, for the first time in ten years, sleeping soundly, with no more watchmen to guard the walls than they would have had in time of peace, were taken by surprise as the Greeks swept through the city, killing, plundering, and burning.

King Priam was slain on the steps of his own palace; Odysseus and Menelaus killed Deïphobus; Hector's little son, Astyanax, snatched from his mother's arms, was flung to his death from the walls, lest he should live to avenge his father. By morning, there were few Trojan warriors left alive; though Aeneas, second only to Hector in battle, escaped and fled from the city.

Now that there was no one left to defend Troy, the Greeks tore down the walls and divided the spoils between them: all the great wealth from Priam's treasure chamber, and the women of his household.

Queen Hecuba fell to the lot of Odysseus, but she did not long survive her husband and her children. Unhappy Andromache was claimed by Neoptolemus, and so she found herself a slave to the son of the man who had killed her beloved husband. He took her with him to his grandfather, King Peleus, and treated her well enough. Old Phoenix sailed with them, but he did not live to reach Phthia, for he died on the way home.

Helen, dragged from the house of Deïphobus by her husband Menelaus, expected no mercy. But Menelaus was ever generous and warm hearted, and she was still the most beautiful woman in all the world, and so he forgave her, and they returned to Sparta together, where they lived in accord and happiness for many years.

With the war over and the spoils shared out, the Greeks set sail for home, but the wrath of Aphrodite still followed many of them. Ajax of Locris was drowned on the voyage, and Diomedes came to Greece to find his kingdom lost to him; but he won himself another kingdom, and lived to an old age.

When Teucer reached Salamis, his father, King Telamon, cast him forth, for he had come home alone, without his brother Ajax.

Agamemnon returned to Mycenae, taking with him great riches, and Cassandra, Priam's daughter; and there, in his splendid palace, he was murdered by his wife Clytemnestra, and by Aegisthus, his cousin, with whom he had inherited a family feud; and Cassandra was killed with him.

Idomeneus was shipwrecked on his way to Crete, and when, at last, he came home, it was to find a usurper ruling his rich kingdom.

Odysseus returned to his little kingdom on the island of Ithaca, but only after ten long years of wandering, and many strange adventures.

Wise Nestor alone, who was much loved by the gods, reached home after a short and fair voyage and lived, respected by all, to a very old age in sandy Pylos, with his many sons about him. Yet he, too, had his sorrow, for he had left his eldest son, Antilochus, dead in the land of Troy.

So ended the long war of the Greeks and the Trojans, a war which, in the end, brought as little good to the victors as to the vanquished.

Names Mentioned in this Book

ACHILLES: son of Peleus, king of Phthia, and of the sea-goddess Thetis; leader of the Myrmidons.

ADAMAS: son of Asius; a Trojan ally.

ADRASTUS: a young Trojan captured by Menelaus and killed by Agamemnon.

AEGEAN SEA: that part of the Mediterranean lying between Greece and Asia Minor.

AEGISTHUS: cousin of Agamemnon.

AENEAS: a Trojan lord; a famed warrior.

AGAMEMNON: son of Atreus; ruler of Argolis and high king of Greece.

AGENOR: a Trojan warrior; son of Antenor and his wife, Theano.

AJAX (the Great): son of Telamon, king of Salamis.

AJAX (the Less): son of Oïleus, king of Locris.

ALCIMEDON: a Myrmidon warrior.

ANDROMACHE: daughter of the king of Thebe; wife of Hector.

ANTENOR: a Trojan lord; one of Priam's wisest counsellors.

ANTILOCHUS: son of Nestor, king of Pylos.

APHRODITE: the goddess of love and beauty.

APOLLO: the Bright One; god of art and learning.

ARES: the god of war.

ARETUS: a young Trojan warrior killed by Automedon.

ARGOLIS or ARGOS: Agamemnon's kingdom in southern Greece; its capital was Mycenae.

ASIUS: a Trojan ally; father of Adamas.

ASTYANAX: son of Hector and his wife, Andromache.

ATHENE: the goddess of wisdom and craftsmanship.

ATHENS: a city in Attica, in northern Greece.

ATREUS: father of Agamemnon and Menelaus.

AUTOMEDON: Achilles' charioteer.

BALIUS: one of the two immortal horses given to Peleus by the gods; the other was Xanthus.

BELLEROPHON: king of Lycia; father of Hippolochus and grandfather of Glaucus.

BRISEÏS: a young woman taken captive by Achilles at the sack of Lyrnessus.

CALCHAS: prophet and soothsayer with the Greek army.

CARIA: a land in south-west Asia Minor.

CARIANS: the people of Caria; under their leader, Nastes, they were allies of the Trojans.

CASSANDRA: daughter of Priam, king of Troy, and Hecuba, his queen.

CEBRIONES: a son of King Priam by a secondary wife; half-brother to Lycaon and Polydorus and to Queen Hecuba's sons and daughters.

CENTAURS: wild and savage creatures, half man and half horse, who lived on Mount Pelion in Thessaly; they were defeated in battle by Theseus and his allies.

CHARIS: a goddess; wife of Hephaestus.

CHROMIUS: a young Trojan warrior.

CHRYSEÏS: daughter of Chryses; taken captive at the sack of Thebe.

CHRYSES: high priest of Apollo; father of Chryseïs.

CLYTEMNESTRA: daughter of Tyndareus, king of Sparta, and sister to Helen; wife of Agamemnon.

CRETE: a large and important island lying south-east of Greece.

DEÏDAMEIA: daughter of Lycomedes, king of Scyros; mother of Achilles' son, Neoptolemus.

DEÏPHOBUS: son of Priam, king of Troy, and Hecuba, his queen.

DIOMEDES: son of Tydeus; king of the cities of Tiryns and Argos, in southern Greece.

DOLON: a Trojan sent by Hector to spy upon the Greek camp.

DORIS: a sea-goddess; wife of Nereus and mother of Thetis.

ERIS: the goddess of strife and discord.

EURYBATES: one of the heralds with the Greek army.

EURYPYLUS: king of the lands lying around Mount Titanus in Thessaly, in northern Greece.

GLAUCUS: son of Hippolochus and grandson of Bellerophon; together with Sarpedon, he ruled over the kingdom of Lycia, in Asia Minor.

GORGON: a frightful monster with the head of a woman, but with serpents for hair.

GREEKS: here, the name used to describe all the inhabitants of the land of Greece, from Thessaly in the north to Sparta in the south, and including all the islands off the mainland.

HADES: brother of Zeus; god of the underworld and of the dead.

HECTOR: eldest son of Priam, king of Troy, and Hecuba, his queen.

HECUBA: wife and queen of King Priam of Troy.

HELEN: daughter of Tyndareus, king of Sparta, and wife of Menelaus; the most beautiful woman in the world.

HELENUS: son of Priam, king of Troy, and Hecuba, his queen.

HEPHAESTUS: god of fire; the smith and craftsman of the gods.

HERA: wife of Zeus; the queen of the gods.

HERMES: the messenger and herald of the gods.

HIPPOLOCHUS: father of Glaucus of Lycia.

IDA: a mountain range in Asia Minor, about 40 miles south-east of Troy.

IDAEUS: the herald of the Trojans.

IDOMENEUS: king of Crete.

ILUS: founder of the city of Troy; grandfather of Priam; his burial mound, surmounted by a pillar, lay outside the walls of the city.

ITHACA: a small island off the west coast of Greece; Odysseus was the king of Ithaca.

LEMNOS: an island lying in the Aegean Sea between Greece and Asia Minor.

LESBOS: a large island lying off the coast of Asia Minor; it was sacked by the Greeks on their way to Troy.

LOCRIS: a kingdom in northern Greece, home of Ajax (the Less).

LYCAON: son of Priam, king of Troy, by Laothoe, a secondary wife; his younger brother was Polydorus.

LYCIA: a kingdom on the southern coast of Asia Minor, ruled by two kings, Sarpedon and Glaucus.

LYCIANS: the people of Lycia; they were allies of the Trojans.

LYCOMEDES: king of the island of Scyros, off the east coast of northern Greece.

LYRNESSUS: a town in Asia Minor, about 60 miles south-east of Troy; the home of Briseïs.

MACHAON: a physician; he and his brother, Podaleirius, were surgeons to the Greek army.

MENELAUS: king of Sparta; son of Atreus and brother to Agamemnon; husband of Helen.

MENOETIUS: ruler of the town of Opus in northern Greece; father of Patroclus.

MERIONES: son of Molus; a Cretan lord; the friend of King Idomeneus of Crete.

MOLUS: a Cretan prince; father of Meriones.

MYCENAE: the capital of Agamemnon's kingdom of Argolis in southern Greece.

MYRMIDONS: the men of Phthia; led by Achilles.

NASTES: leader of the Carians, allies of the Trojans from south-west Asia Minor.

NEOPTOLEMUS: the son of Achilles and Deïdameia.

NEREUS: a sea-god; husband of Doris and father of Thetis.

NESTOR: king of Pylos, on the west coast of southern Greece; father of Antilochus.

OCEANUS: the god of the river, and the river itself, which was believed to encircle the earth.

ODYSSEUS: king of the small island of Ithaca, off the west coast of Greece.

OENEUS: father of Tydeus and grandfather of Diomedes.

OLD MAN OF THE SEA: Proteus, a sea-god.

OLYMPUS: a mountain range in the very north of Thessaly, separating Greece and Macedonia; here, on the highest peak, was the home of Zeus and of all the gods of the upper world.

OPUS: a town in northern Greece, ruled by Menoetius, father of Patroclus.

OTHRYONEUS: an ally of the Trojans, betrothed to Priam's daughter, Cassandra.

PAEONIA: a country lying north of Greece, beyond Macedonia.

PAEONIANS: the people of Paeonia; they were allies of the Trojans.

PANDARUS: an ally of the Trojans, a famed archer, from Zeleia, near Mount Ida.

PARIS: son of Priam, king of Troy, and Hecuba, his queen.

PATROCLUS: son of Menoetius of Opus; the friend of Achilles.

PEDASUS: a horse belonging to Achilles.

PEIRITHOÜS: king of the Lapithae, a people of Thessaly; the friend of Theseus, king of Athens, with whom he fought against the Centaurs.

PELEUS: king of Phthia in northern Greece and ruler of the Myrmidons; husband of the sea-goddess Thetis and father of Achilles.

PELION: a mountain in Thessaly, in northern Greece.

PENELOPE: wife of Odysseus, king of Ithaca.

PERGAMUS: the citadel of Troy.

PHOENIX: Achilles' tutor.

PHTHIA: a small kingdom in Thessaly, in northern Greece, ruled by Peleus, the father of Achilles.

PHYLACE: a town in Thessaly, in northern Greece, ruled by Protesilaus.

PODALEIRIUS: a physician; he and his brother, Machaon, were surgeons to the Greek army.

POLYDAMAS: a Trojan warrior, respected for his wisdom and good counsel.

POLYDORUS: the youngest son of Priam, king of Troy; his mother was Laothoe; Lycaon was his elder brother.

PRIAM: king of Troy.

PROTESILAUS: king of Phylace in Thessaly; the first Greek to land on the Trojan shore, and the first to be killed.

PROTEUS: the Old Man of the Sea; a sea-god.

PYLOS: Nestor's kingdom on the west coast of southern Greece.

RHESUS: king of Thrace; an ally of the Trojans.

SALAMIS: an island lying off the west coast of Attica, in northern Greece; ruled by Telamon, father of Ajax and Teucer.

SARPEDON: an ally of the Trojans; together with Glaucus, he was king of Lycia on the southern coast of Asia Minor.

SCAEAN GATE: the main gateway of the city of Troy.

SCAMANDER: the river flowing through the Trojan plain, and the god of that river.

SCYROS: an island in the Aegean Sea, off the east coast of northern Greece.

SIDON: a city in Phoenicia, famed for its commerce.

SPARTA: a kingdom in southern Greece, ruled by Menelaus.

SPERCHEIUS: a river in Phthia.

STHENELUS: a nobleman of Argos; the friend of Diomedes.

STYX: the river which flowed around the underworld; the spirits of the dead had to cross over it to reach the land of Hades; but no one for whom funeral rites had not been held might cross it.

TALTHYBIUS: one of the heralds with the Greek army.

TELAMON: king of the island of Salamis; father of Ajax and Teucer.

TENEDOS: a small island in the Aegean Sea, a few miles off the coast of Troy.

TEUCER: son of Telamon, king of Salamis; half-brother to Ajax.

THEANO: wife of Antenor, Priam's counsellor; priestess of Athene in Troy.

THEBE: a town on the coast of Asia Minor, south of Troy, sacked by the Greeks.

THESEUS: king of Athens; he fought against the Centaurs with Peirithoüs, his friend.

THESSALY: a district of northern Greece, consisting of a wide plain shut in by high mountains; Mount Olympus is in Thessaly.

THETIS: a sea-goddess; daughter of Nereus and Doris; wife of Peleus and mother of Achilles.

THRACE: a country north of Greece and Asia Minor, stretching roughly from the Danube in the north to the Aegean Sea in the south.

THRACIANS: the people of Thrace; under their king, Rhesus, they were allies of the Trojans.

THRASYMELUS: a young follower of Sarpedon, king of Lycia; killed by Patroclus.

TIRYNS: a city in Argolis in southern Greece, ruled by Diomedes.

TITANUS: a mountain range in Thessaly in northern Greece, home of King Eurypylus.

TROJANS: the inhabitants of the city of Troy and of the surrounding plain.

TROY: a city in Asia Minor, and the land lying about it.

TYDEUS: father of Diomedes.

TYNDAREUS: king of Sparta; father of Helen and Clytemnestra.

XANTHUS: one of the two immortal horses given to Peleus by the gods; the other was Balius.

ZELEIA: a town at the foot of Mount Ida; home of Pandarus.

ZEUS: the king and father of the gods.